The American Coal Miner

A Report on Community and Living Conditions in the Coalfields

The President's Commission on Coal

John D. Rockefeller IV
Chairman

Washington : 1980

Material used by permission: p. 11, © 1977, North Cambria High School, Barnesboro, Pa.; p. 80, © 1968, Richard Carter, New York.

Library of Congress Cataloging in Publication Data

United States. President's Commission on Coal. The American coal miner.

Bibliography: p.
1. Coal-miners — United States.
I. Rockefeller, John D. IV, 1937-
II. Title.
HD8039.M62U66975 1980
331.7'62233'0973   79-607928
ISBN 0-9603806-1-2

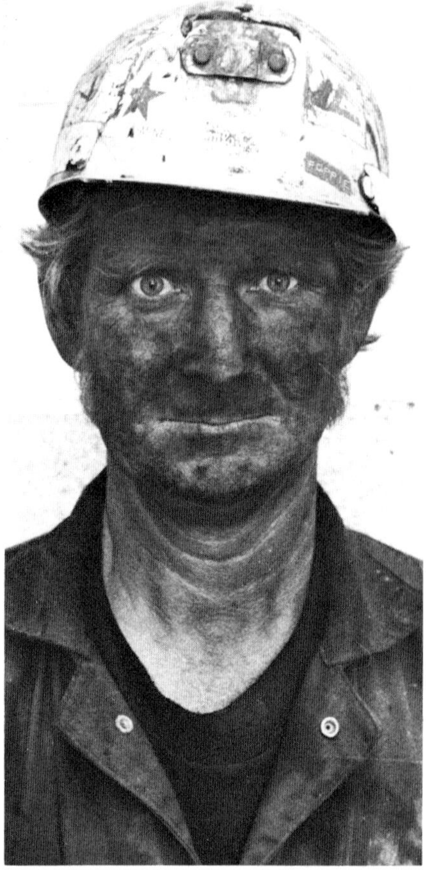

The coal mining work force consists of men and women of diverse backgrounds from many regions of America. (Photographers are identified in parentheses): Col. 1: Top: Alabama (Murray); Center: West Virginia (Murray); Bottom: Kentucky (Wathen). Col. 2: Center: Utah (Corn); Bottom: Oklahoma (Simmons). Col. 3: Top: Kentucky (Wathen); Center: Virginia (Murray); Bottom: Alabama (Murray). Col. 4: Top: New Mexico (Murray); Center: Illinois (Wathen); Bottom: West Virginia (Corn). Col. 5: Top: Kentucky (Corn); Bottom: Ohio (North American Coal Co.). Col. 6: Top: West Virginia (Corn); Center: Kentucky (Corn); Bottom: Virginia (Murray). Cover photo: Webster County, Ky. (Wathen).

Foreword

Coal is unique among the major energy resources in its close identification with the people who produce it.

Unlike oil, natural gas, and hydroelectric power — which are produced with a relatively small and anonymous work force; or nuclear, with its unheralded uranium miners and highly trained nuclear specialists — coal and the coal miner are intertwined in American legend and history.

When we think of the coal miner, various images come to mind. Powerful and defiant union leadership is personified for many by the memory of John L. Lewis. We remember with sadness the tragedy and treachery of the Yablonski murders. And, perhaps most commonly of all, many Americans see the coal miner as poor, downtrodden, and wearied by harsh labors in dangerous and dirty mines — struggling to clothe and feed a family in a rundown house somewhere in an abandoned corner of the country.

The Executive order that created the Commission called for an examination of a variety of issues, including "health, safety and living conditions in the Nation's coal fields." At the first formal meeting of the Commission, on September 29, 1978, this subject was designated as one of the four main areas to be pursued in the work of the Commission.

During the course of four public hearings in the coalfields and visits to surface and underground mines and to the homes of coal miners in the fall of 1978, we heard numerous complaints from coal miners about the image that Americans have of them. They expressed concern about what they saw as a combination of less-than-objective news coverage and a general unwillingness by most Americans to believe that improvements have come to the coalfields.

Because of the compassion and concern that the American people hold for victims of poverty and social injustices, and the close interrelationship of coal and the coal miner in the minds of most Americans, the Commission decided that a report on how modern coal miners live and work would be a valuable contribution to the national energy debate at a time when the role of coal is becoming more important.

This report differs from many others issued by Federal agencies and commissions because so much of our material comes from the miners themselves, nearly 1,300 of whom were interviewed by our field teams.

In terms of the scope of the field work by interviewers and photographers, this report represents the first major documentation of coalfield community and living conditions since the Boone Report of 1947. That report, officially entitled *A Medical Survey of the Bituminous-Coal Industry,* was published by the Coal Mines Administration of the Department of the Interior in 1947 in response to a demand by John L.

Lewis for improvements in public health and medical treatment at the mines and in the coal camps.

The situation in 1979 is very different from what it was in 1947. Not only do the coal miners enjoy a much more comfortable lifestyle and significantly improved quality of health care, but also a much higher percentage of miners are employed at surface mines, many of which are now in the western States.

During the decade of the 1970s, women entered the mining work force, often after successfully seeking legal action under Federal Equal Employment Opportunity statutes. Theirs is a story of a struggle, which continues today, to become full-fledged members of the work force. The number of Native American coal miners has also increased significantly with the opening of large surface mines on the Navajo Reservation in Arizona and New Mexico.

Many problems remain in the coalfields. There is vast room for improvement in the areas of housing and highways. Because of the rural location of many coalfields, health care facilities and services are less extensively available than in more urban areas. Coal mining remains the most dangerous of occupations, despite considerable improvements in coal mine health and safety. (These problems are addressed in the Commission's Findings and Recommendations volume.) And poverty still exists in some coal communities, even though the conditions of deprivation are no longer descriptive of the lifestyles of today's well paid working miner.

The progress achieved by coal miners over their counterparts of earlier generations cannot be denied. Thirty years ago, the miner and his family were still living in coal camps, often in conditions of poverty. Today, mining communities are much more diverse, and the modern miner is, more often than not, among the resident middle class.

The reality of that vast change is not widely recognized. All Americans should take pride in this success story, and should understand that the miners stand ready to do their part in assuring that the Nation has an adequate supply of its most abundant energy resource.

This book concentrates on employed miners throughout the country, a widely diverse group of men and women whose living conditions are as different as are Appalachian hollows from western plains. *The American Coal Miner* is the story—as best we could tell it, relying heavily on the words of the miners themselves—of a proud group of American workers toward whom an energy-dependent Nation now turns.

John D. Rockefeller IV
Chairman

Charleston, West Virginia
February 1980

Introduction

The American Coal Miner is an account of how coal miners live and of how their lives have changed over the past 30 years. It weaves data, personal interviews, government reports, historical and current literature, and formal surveys to depict contemporary living conditions and daily life in coal-mining communities from Appalachia to the West.

Executive Order No. 12062 charged the President's Commission on Coal with conducting "a comprehensive review of ... health, safety and living conditions in the Nation's coal fields."

Two earlier major studies of living conditions in the coalfields have been undertaken. The first was the massive statistical study of the U.S. Coal Commission in 1922 that produced 5 volumes of text and tables containing a detailed and intricate examination of life in major coalfields in the East and Midwest.

The second study, *A Medical Survey of the Bituminous-Coal Industry,* is commonly referred to as the Boone Report after its director, Rear Adm. Joel T. Boone. That study was conducted in 1946 by the Coal Mines Administration, then part of the Department of the Interior. It concentrated on gathering statistical data from site visits in coalfields in the East and the Midwest, as did the 1922 Survey, but also covered mining communities in six western States.

The President's Commission on Coal, aware of the contributions made by the two previous studies and desiring to achieve comparability where possible, decided to adopt the most promising methodology used in them and to add features of its own. Like the earlier studies of coal community life, the Commission developed a detailed questionnaire about miners' lives and living conditions. The addition involved face-to-face interviews with miners, their families, local public officials, and members of the business community.

The interviews allowed us to go beyond the limited information any questionnaire can provide. For example, miners expressing general satisfaction with their housing in the questionnaire responses often spoke eloquently and at length during interviews about the frustrations of providing a family with adequate housing, particularly in central Appalachia.

Like the 1946 Report, the Coal Commission made extensive use of photography to capture the reality of living conditions today.

In spring 1979, the Coal Commission sent 15 experienced interviewers, journalists, and researchers along with four photographers through the coalfields to talk with miners and their families. The interviewers developed much of the data on which the report relies. The questionnaire was administered to randomly selected miners at 55 mine sites in 17 States.

Independent research organizations provided additional information from current and historical data and a survey on living conditions in coal-producing areas.

Interviewers traveled in three-person teams throughout five coal-mining regions designated by the Commission: northern Appalachia (Pennsylvania, Ohio, and northern West Virginia), central Appalachia (southern West Virginia), southern Appalachia (eastern Kentucky and southwestern Virginia), Midwest/Alabama (Alabama, Illinois, Indiana, and western Kentucky), and West (Montana, Wyoming, Colorado, Utah, Arizona and New Mexico).

Over a 7-week period, survey teams met with miners just before or after they finished their work shifts. Interviews were conducted at all hours — such as midnight and 8 a.m. — and in many places — outside bathhouses and in tiny offices at the mine portal. Questionnaires were administered at each site to up to 50 miners who then participated in small group interviews. Interviewers also visited the homes of some of these miners to talk with their families.

Responses to the questionnaire provided facts about the miners' living conditions (types of homes rented or owned, shopping and medical facilities available, income, road conditions), and about their satisfaction with these conditions. The questionnaire was administered to 1,299 miners; the results (based on 1,281 complete questionnaires) are reported in this book as the Living Conditions Survey.

The interview material provided personal detail on living conditions.

As expected, these two methods of gathering information — questionnaire and personal interview — drew different reactions from miners. On the questionnaire, miners indicated that they were generally satisfied with community conditions (except roads, which always elicited complaints). But during interviews and lengthy conversations they talked about the problems in their lives and communities — expensive and hard-to-obtain housing, poor utility service, inadequate schools, spotty medical care, and lack of recreation facilities. (See Appendix, "The Living Conditions Survey," which provides details on the questionnaire and interview process.)

Survey team members telephoned weekly summaries of their interviews to the Commission's offices in Washington, D.C.; by June, the onsite interviewing and information gathering were largely finished. The Commission then assembled a group of writers to begin work on *The American Coal Miner.*

Additional sources of information were developed, drawing on the Library of Congress, Federal agencies, regional organizations, State and county governments, academics, and mining industry and union experts.

One purpose of this book is to chronicle changes in the living conditions of miners since the Boone Report, a document that stunned readers with its account of squalid conditions in post-World War II coal communities.

As is evident to the reader, much has changed since then. The coal miner now demands and generally has access to those amenities and necessities available to other well paid industrial workers. Although some problems still exist, especially with regard to housing and roads in central Appalachia, the real story of the American coal miner is this: even in communities with problems of poverty, the working coal miner has risen to the status of a middle class citizen, and ought not to be regarded as a member of an impoverished class.

The relative status of the miner does not change the fact, of course, that local and regional economies in coal regions remain in need of development assistance to promote diversification and to eliminate historic deficits in public investments.

Fortunately for this effort, two of the Boone Report's major contributors, Russell Lee — who traveled through the coalfields photographing miners and their families — and Allan Sherman — a principal writer of the Boone Report — were available to the Commission to recollect their impressions of coal miners' lives in the early postwar period and to comment on how conditions differ today. Mr. Lee also served as an adviser on the illustration of the book.

The American Coal Miner is not an exhaustive statistical study of coal miners or of the living conditions in the communities where they live. The report does provide, however, a contemporary study of housing, education, medical care, recreation, family life, and other subjects.

This book represents a unique and timely look at a vital part of the energy work force in the United States. As the national debate on energy policy evolves, this book may help many government decisionmakers and members of the public to associate the debate with the work force involved, and to recall that policy ought to account for people's lives as well as fossil fuels, power plants, and pipelines.

Michael S. Koleda
Executive Director

Washington, D.C.
February 1980

Pages vi and vii: Coal-fired generating plants produce 47 percent of the Nation's electricity. Left: Power plant on the West Virginia side of the Ohio River, viewed from Powhatan Point, Ohio. Top right: Surface mine near Morris, Ill. Bottom right: Underground mine in western Kentucky. (Wathen)

The President's Commission on Coal

John D. Rockefeller IV
Governor of West Virginia
Chairman

Jesse F. Core
Adjunct Professor of Mining Engineering
Pennsylvania State University
University Park, Pennsylvania

Marvin Friedman
Vice President
Ruttenberg, Friedman, Kilgallon, Gutchess & Associates, Inc.
Washington, D.C.

W. Dewey Presley
Director and Chairman of the Executive Committee
First International Bancshares, Inc.
Dallas, Texas

W. Willard Wirtz
Wirtz & La Pointe
Washington, D.C.

Ex Officio

Walter D. Huddleston
United States Senator from Kentucky

Charles H. Percy
United States Senator from Illinois

Jennings Randolph
United States Senator from West Virginia

Charles W. Duncan, Jr.*
United States Secretary of Energy

John Buchanan
Member of the House of Representatives from Alabama

Austin J. Murphy
Member of the House of Representatives from Pennsylvania

Carl D. Perkins
Member of the House of Representatives from Kentucky

Ray Marshall
United States Secretary of Labor

Executive Director
Michael S. Koleda

**Succeeded James R. Schlesinger on August 24, 1979*

Coal miners live where coal can be found—in areas as richly diversified as the Nation itself. They live on the fertile farmlands of the Midwest (top), in quiet towns in Appalachia (bottom left), and in expanding coal boom-towns of the West. (Wathen, Corn, Corn)

Table of Contents

Coal and countryside blend in West Virginia (top), where cooling towers at a coal-fired electric generating plant resemble those at nuclear power plants. (Wathen) Bottom: A coal tipple looms over a ranch near Gillette, Wyo. (Murray)

Coal Mining In America

Must
Reset

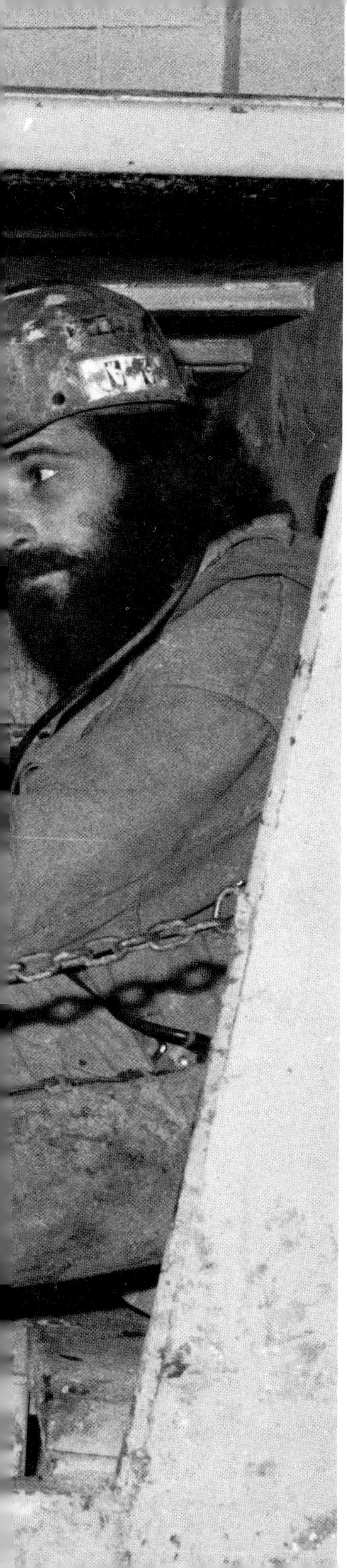

The fortunes of the Nation in the 1980s are linked to coal and to the men and women who mine it.

Coal is the most abundant and readily available energy fuel in the United States. At present rates of use, known reserves will last for at least 300 years.

Rich deposits of coal underlie the hills and hollows of Kentucky and West Virginia. Thick beds lie just below the surface of Arizona, Montana, Wyoming, and other western States. Vast seams are scattered across the Illinois Basin and Pennsylvania. Coal is mined in 26 of the 31 States where reserves have been found.

At a time when the Nation is searching for alternatives to imported oil, coal offers a ready and reliable substitute.

An industrial work force of 233,400 men and women mine coal in the United States today. These miners have become important figures in the growing public debate about national energy policy.

They dig the coal that fires boilers that generate 47 percent of the electricity consumed in the United States. Their work produces coal that, as coke, is used in steelmaking and, as coal tars, is an important raw material in manufacturing chemicals and other industrial products vital to the economy.

The stability and vitality of this work force are key ingredients in the effort to tap the energy resources of the Nation. The miners and their well-being, therefore, are important subjects for consideration in the national energy debate.

Who are the coal miners? Where do they come from? Where and how do they live? How satisfied are they with their housing, transportation, and health care? How comfortable are their living conditions?

The answers to these questions can go far in providing answers to more fundamental questions. How reliable is the coal-mining work force? If the Nation calls upon the coal industry to provide more fuel to meet national energy requirements, to what extent can the miners be expected to respond with the necessary will and work?

This report provides information that can help to assure that the national energy debate — to the extent that it deals with coal miners — rests on an accurate picture of contemporary coal miners.

Overleaf: The Nation has long depended on coal, its most abundant energy resource. Coal from mines such as this one near Helper, Utah, is increasingly being used to replace oil and gas in the generation of electricity. Coal from this mine fuels a power plant less than ½ mile away. (Corn)

The American coal miner plays a vital role in the production of fuel to meet the energy requirements of the Nation. These miners ride a "mantrip" to their workplace deep underground near Washington, Pa. (Wathen)

Old Images of Coal Linger

For many Americans, the mention of coal may conjure up images of a wiry Kentuckian, unaware of the realities of contemporary life, who faces the hardships and dangers of digging coal in poverty-stricken areas of Appalachia.

For others, coal may evoke memories of the rich history of mining, a history captured in the literature of Dickens, D. H. Lawrence, and Zola, in the verse and song of the Welsh miners, and in vivid accounts of mining in this Nation.

Coal may evoke other responses. For many, the image of coal conjures up the grief-stricken faces of miners' families at the site of a mine explosion: Monongah, Castle Gate, Centralia, Farmington, Scotia—the list is long.

For some, coal means harsh working conditions imposed by hardnosed and insensitive operators with their iron and coal police and their scabs.

For others, coal means the wildcat strikes and the brawling of the miners, indifferent, some say, to the national interest, and concerned only with their own welfare. Did they not strike during World War II?

Then there was John L. Lewis, a legendary son and grandson of coal miners, who rose from the coalfields of Illinois to become one of the most powerful figures on the American labor and political scenes from the 1920s to the 1950s. He helped to build the United Mine Workers of America (UMWA) into a formidable political power on behalf of the miners—and he earned much public and private enmity in doing it.

When Lewis retired in 1959, after almost 40 years of service to the union, he was flooded with letters of praise from the powerful and the poor. One miner wrote:

My little local 6374 in Va. has been in action since 1933 June, which you will remember up to that time we were in Slavery *in the Va. coal mines....*

Now for me and my wife we want to thank you *from our hearts for the Hospitals, Pensions, and for bringing us out of slavery. I guess the Welfare fund has spent* six or seven *thousand dollars on Mrs. Wallen and about $800.00 on me....*

Now I know there is no miner any more thankful than I am for our organization in this U.S.A.

Now Mr. Lewis we hope you the very best of health and many more years to stand by us.

More recently there was the tragic story of Joseph A. "Jock" Yablonski, the union official who was leading a rank-and-file rebellion against the entrenched and autocratic leaders of the UMWA, only to be murdered in 1969 with his wife and daughter in their beds at home in southwestern Pennsylvania. Yablonski's foe, incumbent UMWA Presi-

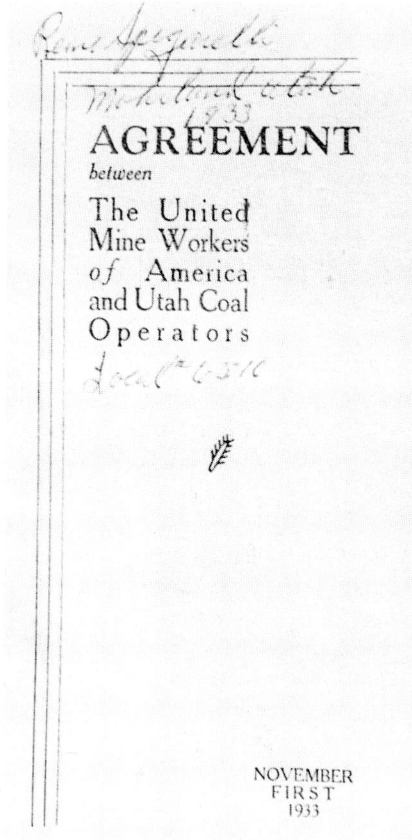

AGREEMENT
between
The United Mine Workers of America and Utah Coal Operators

NOVEMBER FIRST 1933

dent William A. "Tony" Boyle, was convicted of conspiracy in the murder plot.

Yet the image of the miner that may remain today is not that of a Welsh miner, or a contentious Lewis, or a martyred Yablonski. Rather, it more likely is that of a downtrodden worker

For decades, coal miners and coal operators have grappled over issues such as wages, hours, health care, and safety conditions. Sometimes strikes and lockouts occurred. Over the long term, however, pay and working conditions for miners have improved enormously. Top left: State police escort nonunion workers through a picket line in Harlan, Ky., long the scene of labor unrest among miners. Bottom: Lois Scott of Harlan, a miner's wife, holds a portrait of John L. Lewis, for many years the president of the United Mine Workers of America. (Corn) Top right: A 1933 labor agreement from Utah. (Corn)

whose life grinds along on the edge of poverty and injury, whose wife is a gaunt and craggy figure surrounded by dirty-faced, ill-clothed children.

That image doubtless was accurate 30 or 40 years ago, when coal companies operated company towns and maintained a kind of feudal relationship with their workers. Mining was a low-paying occupation, and miners were part of a vast pool of semiskilled labor, a pool too often considered all but expendable by some operators and by some elements in government, neither of whom seemed to be sufficiently concerned about the dangers in mining.

That picture of the coal miner was reinforced in the early 1960s by national political developments. In the spring of 1960, with the coal market in a slump and many miners out of work, two Presidential hopefuls — John F. Kennedy and Hubert H. Humphrey — engaged in a fateful struggle in the Democratic primary election in West Virginia. Their campaigns highlighted the plight of out-of-work miners, living in substandard conditions, often wanting for the basic elements of sustenance.

This picture of coal miners in Appalachia moved government leaders and voters alike, finally giving birth to both the idea and the rhetoric of the War on Poverty.

That image no longer represents reality, at least not for the men and women who mine coal. Some problems in the coalfields remain. There are pockets of poverty among the unskilled and the elderly of Appalachia. The quality of health care and public education is substandard in some coal regions. Chronic shortages of adequate housing and poor roads create problems for everyone living in that region.

But there have been enormous changes, mostly in the direction of a better life for the miners.

The New Miner

The American coal miner today has little in common with the stereotypes of 30 years ago. The working miner today is far better paid than the miner of the 1940s, in relative as well as absolute terms. Miners' incomes in 1979 rank near the top of those of all industrial workers.

Miners today are better educated and better trained than their counterparts of a generation or two ago. They have a vastly improved system of health care. They own new cars, furnish their homes well, take vacations in various parts of the Nation, and in many other ways mirror their colleagues in the auto, steel, or other industries.

Retired miners depend on their union pension, social security, and (for some) Black Lung benefits for subsistence. Currently, 73,896 miners are carried on the UMWA retirement and pension rolls. Top left: A meeting of UMWA retirees in Helper, Utah. (Corn) Bottom left: John Popp of Howard, Pa., retired after 40 years in the mines. (Wathen) Top right: Charles D. Hodge, a miner retired on Black Lung disability, now the mayor of Gary, W. Va. (Corn) Bottom right; Sam Fazzio of Helper, Utah, born in Italy and a coal miner in Utah since 1920, with a picture of himself in 1922. (Corn)

When Coal Became King

Coal has always been available when the Nation required a cheap, plentiful source of energy.

Wood was the primary fuel for heating homes and fueling the furnaces of the fledgling industries of the Nation until sometime after the 1840s. Coal had been mined commercially in the new country since 1750, but it was used mostly as a substitute for wood for home heating. By the middle of the 19th century, a number of small mines produced coal for such local consumption.

Then, in the second half of the 1880s, came the rapid expansion of industry and of the railroad system. With the advent of the coal-fired steam engine, the railroads became both the major consumers and the major transporters of coal, hauling it from the rich coalfields of Appalachia and Illinois to the burgeoning cities in the Midwest and the East. New processes permitted the rapid conversion of coal into coke that, mixed with the high-grade iron ore from the Mesabi Range in Minnesota, soon turned the United States into a steel-making giant.

The rise in coal production created the great coal companies — Peabody, Consolidation, Island Creek, Pittston — that dominate the field today. Many coal operations have been bought in recent years by oil companies; mining, insurance, steel, public utility, and construction firms also have bought into this promising energy field.

Coal has generated its share of "barons" of the industry — the men who made countless millions — as well as a number of self-made men, the hallmark of American industry of the period. One of these was William Leckie.

Born in Scotland in 1855, he entered the mines when he was 10, emigrated to the United States at 21, worked in the anthracite fields of Pennsylvania, saved his money and earned a

bachelor's degree at Dickenson College, worked up through the ranks of several companies, began acquiring coal interests at 51, and soon owned or sat on the boards of a number of coal and land companies (including some he had worked for years before) and a bank. He located his operations in Welch, W. Va., and prided himself on his coal camps, which "had modern homes, concrete walks and landscaping."

Then there was Francis Stuyvesant Peabody, Chairman of the Board of Directors of the Peabody Coal Co., a giant in the industry since early in this century and the largest producer of coal in the Nation today. He was born in 1859, graduated from Yale University, entered the coal business, and built the company that by 1918 had operations ranging from Virginia to Wyoming.

While the Leckies and the Peabodys managed the affairs of coal companies, the men in the mines lived in a different world.

One of those miners was John Kolonay, who immigrated from Austria-Hungary (now Czechoslovakia) to the United States in 1913 and settled in Frugality, Pa. There he married Elizabeth Sipos; they would have 12 children. A granddaughter recently wrote this account:

Underground mining has become highly mechanized and is safer than in the past. Top left: Miners leaving the mantrip for the walk to their work stations in a mine near Webster, Ky. Bottom: A shuttle car driver in a mine near Washington, Pa., receives a load of coal from a continuous miner, which cuts coal from the face of the seam and loads it in a continuous operation. Top right: In the same mine, a miner installs roofbolts, which are long metal rods used to keep the mine top from falling. (Wathen)

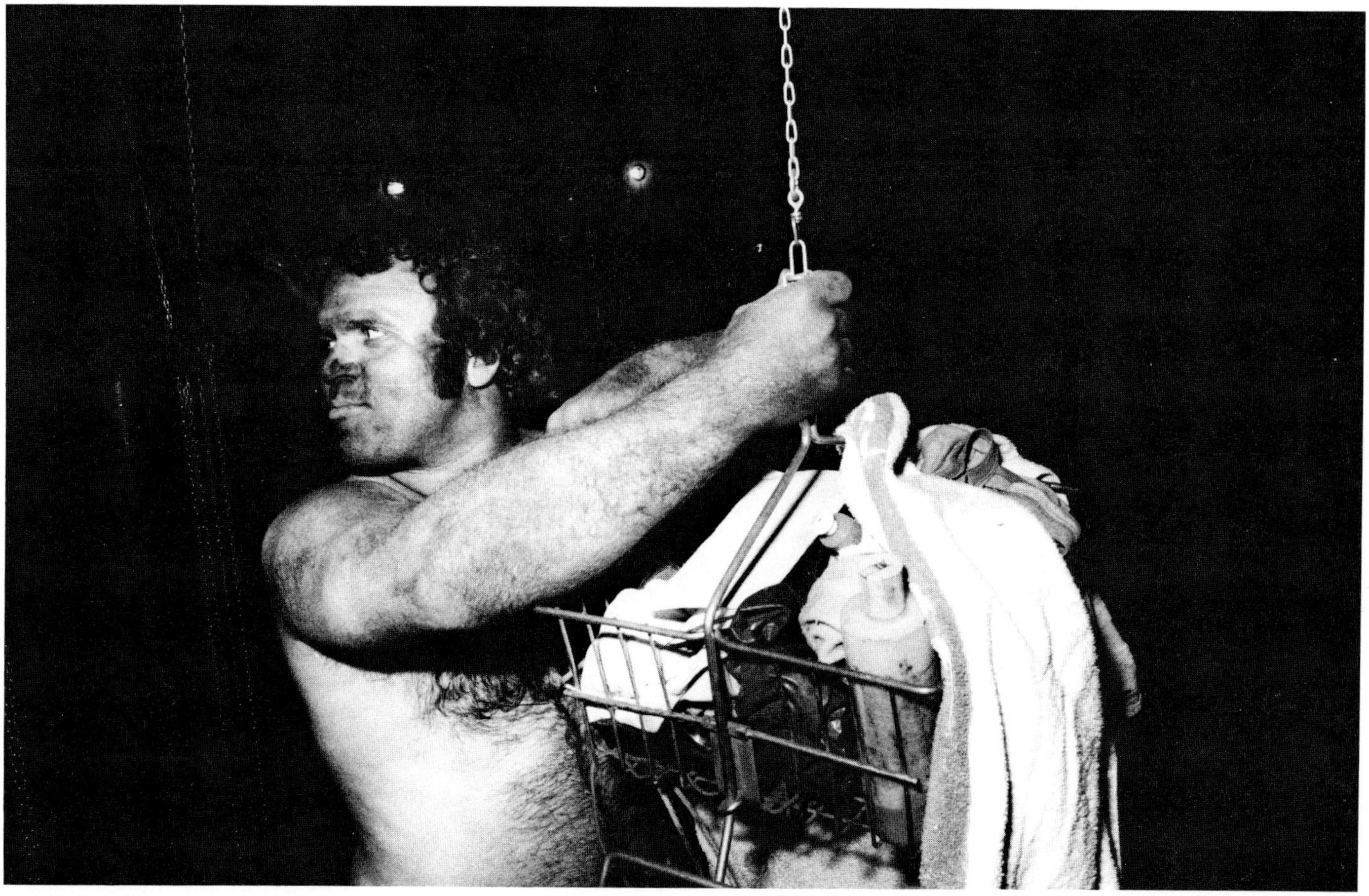

Grandfather would sometimes walk to work at 4 a.m. with all his miner-neighbors. Because it was still dark, only their carbide lamps could be seen moving down the road. Grandfather carried his lunch — some coffee-flavored water, sandwiches, and sour pickles to get the taste of coal out of his mouth.

Working for long hours on his knees with pick and shovel and breathing coal dust left Grandfather very tired. He would come home coal black. My grandmother would set up the tin tub by the coal stove so he would be warm while she washed his back with Octagon soap. His clothes, always wet and filthy, often had to be sewed and dried over the back of the coal stove.

Grandfather Kolonay worked with black lung for many years; still, it was very hard for him to receive benefits. He retired from the Rich Hill Mine at 64 after 46 years filled with the danger and uncertainty that each day underground brought. John Kolonay's story is the story of thousands of immigrant miners who measured up to the tests of time.

By the early 1900s and well into this century, coal was king of the U.S. energy industry. Total production finally peaked at 631 million tons in 1947. Following World War II, in which coal played a vital role, production seesawed down to a bottom of 363 million tons in 1954 and another low of 366 million tons in 1959; the market was depressed throughout the early 1960s, but then came to life again as demand rose in the 1970s.

The industry produced about 770 million tons in 1979, a level that Coal Commission projections foresee as more than doubling by 1990 and possibly tripling by the year 2000.

Until recently, coal inevitably produced serious environmental problems. For decades, photographs, cartoons, and stock certificates depicted tall stacks belching black smoke and boasted of "progress" in the new Nation. Little attention was paid to the human price, and only well into the 20th century did medical and scientific evidence accumulate on the dangers of polluted air.

Some cities moved comparatively quickly to clean up both their air and their image. Notable among these was Pittsburgh,

Underground mining has always been hard work, but modern miners — unlike many of their predecessors — take hot showers and change into casual clothes before leaving for home. Left: Miners run workclothes up on lines that they lock when leaving this bathhouse at a mine near Powhatan Point, Ohio. Right: Miner John J. Heltsley finishing a shift at a mine near Webster, Ky. (Wathen)

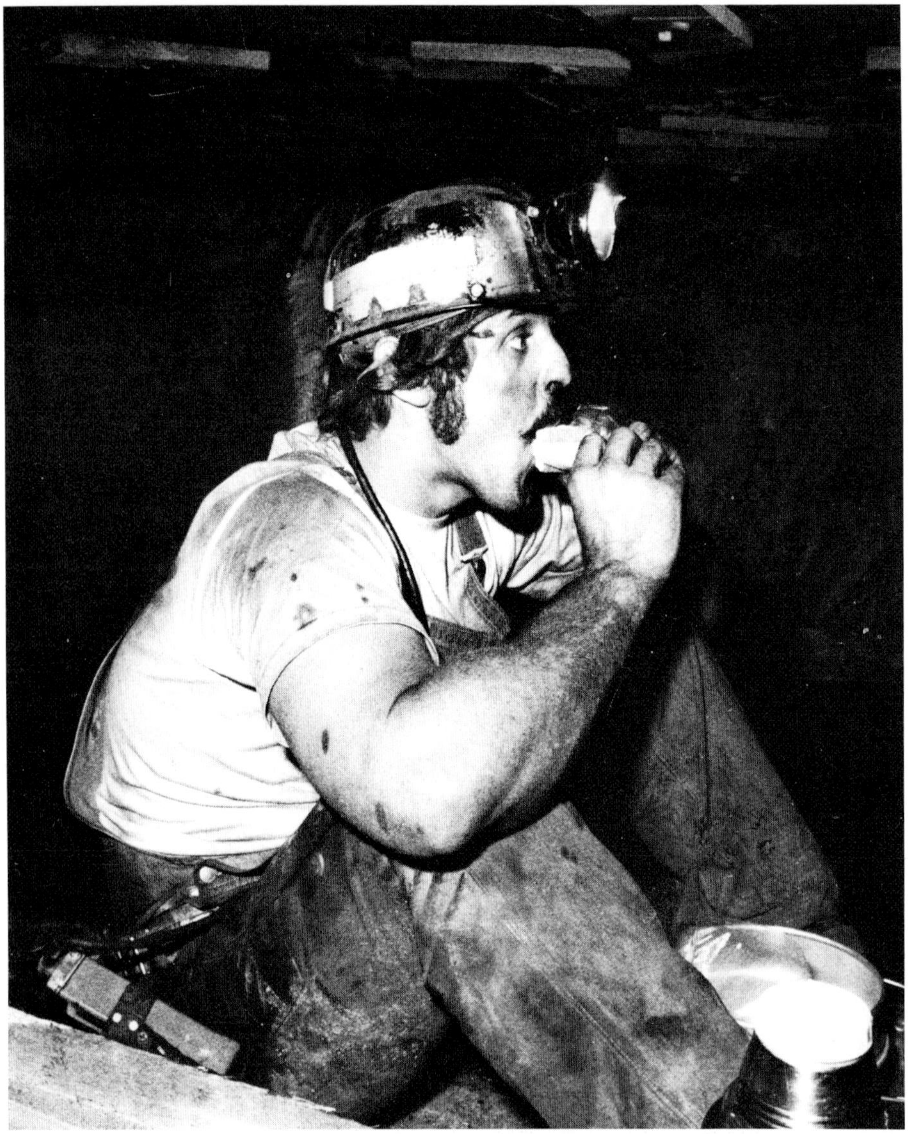

Pa., which had perhaps the worst air of any major industrial city in the Nation in the 1930s and 1940s. With a vigorous effort following World War II, the city virtually eliminated the noxious effects of sulphur dioxide emissions from the homes and industries. The image of Pittsburgh as a city of dirty air lingered afterward, but that, too, finally was dispelled.

Today, coal-fired generating plants can be operated at least as cleanly as those fired by oil. In many cases, with modern "scrubbers" and precipitators, coal-fired plants actually are cleaner in the amount of solid particulates and sulfur dioxide emitted than are the oil-fired plants they replace. Dirty air from coal need be no more than a memory from the past.

Coal long was associated with mining disasters, which almost always resulted from explosions in deep underground mines.

Those tragedies finally prompted national action, culminating in the Federal Coal Mine Health and Safety Act of 1969, which imposed strong requirements on coal-mining operators to observe stringent safety regulations. Largely as a result of this legislation and the ensuing work of the Mine Safety and Health Administration, coal mining today is far less susceptible to explosion than it was even a decade ago. The most recent serious underground explosion was in 1976, when 23 miners and three Federal inspectors died in Scotia, Ky.

Improvements in mining engineering, safety procedures, and rescue techniques continue to reduce the risk of disasters in coal mining. The rate of fatalities in coal mining is now one-eighth of what it was 35 years ago. But many miners died under the working conditions, which sometimes were in blatant disregard of minimum standards of mining safety, that existed in those bygone times.

Company Men and Union Men

The coal industry consists of many people — coal mine operators, miners, transporters, users — but it is the mine owners and the union members who have dominated the public scene.

The battle between companies and unions has been as hard-fought, as tragic, as colorful, and ultimately as successful for both sides as any in the history of American business and labor.

From about 1750 to the 1890s, the companies faced no organized demands from the miners, who were largely illiterate, poorly paid, and politically powerless. Working conditions in many mines were bad: most mine operators concentrated on profit and paid scant attention to safety or health of their men. Major disasters from explosions occurred regularly.

As coal mining grew in importance after the Civil War and mining operations increased, miner resentment against the operators also increased, spawning the beginnings of the effort to organize the coalfields and giving rise to radical groups, the most notorious of which was the Molly Maguires.

The Molly Maguires were supposedly a secret society of Irish miners, existing from 1850 to 1877. They objected to the living and working conditions in the anthracite fields of Pennsyl-

vania, where they were especially strong. They contended that coal company camps were unfit for living or raising a family in, that pay was too low, and that the mines were unhealthy and unsafe.

Using terrorism in addition to political strategies, the Molly Maguires burned down coal tipples, blew up locomotives, and even shot company supervisors. The coal operators eventually hired a Pinkerton agent to infiltrate the secret society; upon his testimony, 20 miners in 1877 were tried for violent acts, convicted, and hanged.

In those same years, the foundations were laid for what has become the major coal-mining labor union in the United States — the UMWA.

The union was founded in 1890 and drew its early strength from miners in the Midwest. Its active U.S. membership in January 1979 was 184,539 nationwide, but it remains particularly strong in the Midwest and the East. There were also an estimated 41,000 nonunion miners, mostly in the West but also in mines scattered throughout other coalfields.

Union miners now produce about half of the coal mined in the United States — the UMWA share is about 48 percent — illustrating the trend in recent years toward increased production by nonunion labor. The reason appears to lie in the growing production of large, nonunion surface mines in the West. The productivity per miner in surface mines exceeds that in underground mines by a substantial margin. Some miners, especially in the western surface

Left: Time for a break in the "dinner hole," where miners eat their mid-shift meal. This miner is in a mine near Webster, Ky. (Wathen)

Right: Surface mining depends on huge draglines or shovels that remove the overburden, which ranges from a few inches to many feet thick. The overburden is carefully preserved and is spread back in place. This mountainside near Abingdon, Va. has been contoured and seeded so that it no longer bears the marks of mining. (U.S. Dept. of the Interior)

mines, belong to unions other than the UMWA; these include the International Union of Operating Engineers, the Progressive Mine Workers, and the Southern Labor Union, among others.

The coal operators also have an organization — the Bituminous Coal Operators' Association (BCOA), which consists of more than 130 operators in the Nation. Both the UMWA and the BCOA maintain headquarters in Washington, D.C.

Strife, Strikes, and Studies

Relations between the coal operators and the union were seldom peaceful. The issues that divided the two sides were often important: fair pay, decent working hours (miners once worked 12 and 14 hours a day), child labor (boys once worked in mines), health care and safety in the mines, and living conditions in the coalfields.

But the road to resolving disagreements between the company men and the union men was turbulent indeed. The companies often seemed callous; they operated unsafe mines, often locked the miners out, and gained a reputation for insensitivity and greed.

The miners, for their part, called major strikes that shut down vast segments of the economy: As coal stockpiles dwindled and disappeared, the steel indus-

try, the railroads, the automobile industry — the very heart of the national economy — would grind to a halt. More recently, the miners gained a reputation for calling wildcat strikes, sometimes for seemingly trivial reasons.

One such strike occurred in 1922, and led Congress to create the United States Coal Commission that year. The U.S. Coal Commission conducted an exhaustive study of living and working conditions in the coalfields. The study concentrated on the economic aspects of the industry and the work force, and it explored such subjects as illiteracy rates among miners, cost of living in the coalfields, and sanitation.

Surface mining produces an increasing share of the Nation's coal — an estimated 59 percent of the total produced in 1979. Top left: Operator's view of a dragline, this one run by Willard G. France in a mine in Fulton County, Ill. (Wathen) Bottom left: Mr. France controls the huge machine with levers and buttons. (Wathen) Top center: A large dragline operating near Cumberland, Ohio. (Wathen) Bottom right: After the overburden is removed, a loader scoops up coal and dumps it into a truck in Harlan, Ky. (Corn) Top right: Sign along a Wyoming highway. (Utrup)

The Boone Report

The report of the U.S. Coal Commission remained the most important Federal study of the coalfields for 25 years. In the closing months of World War II, another major miners' strike led to a second major Federal study —and to a turning point in the history of coal mining in this Nation.

John L. Lewis called the strike at the end of particularly rancorous negotiations with the coal operators. President Harry S. Truman was incensed, and ordered the Federal Government to seize the mines in the interest of national security. The miners promptly returned to work under the administration of the Government. The miners worked under an agreement—the Krug-Lewis Agreement—reached between Lewis and Secretary of the Interior J. A. Krug.

The Krug-Lewis Agreement contained provisions that changed the health care system provided to the miners and opened a new era in labor-management relations in the industry. These provisions established a welfare and retirement fund, financed by a company-paid royalty of 5¢ per ton of coal mined, and a medical and health fund financed by money deducted from the miners' paychecks.

When the operators and the UMWA eventually signed a new contract, the provisions of the Krug-Lewis Agreement were incorporated, and thus began the Health and Welfare Fund of the UMWA, which became a model in the provision of health care. A fuller account of these developments is presented in Chapter 3.

Another section of the Agreement called for a Federal study of health and medical care in the coalfields. Secretary Krug entrusted this task to Rear Adm. Joel T. Boone, a Navy doctor who had won a Congressional Medal of Honor for heroism in World War I, who had been Physician to three Presidents, and who was widely admired in the medical profession.

Admiral Boone sent teams of Navy doctors and technicians into the coalfields to measure the availability and effectiveness of the health care for miners. He retained the celebrated photographer, Russell Lee, to illustrate the study. The effort resulted in a massive work, *A Medical Survey of the Bituminous-Coal Industry,* published in 1947.

The report, referred to in this book as the Boone Report, gave the Nation information and the pictures that chronicled the poverty, poor health, and substandard living conditions of the miners.

The Boone Report remained the last major Federal study of living conditions in the Nation's

Surface mining in the hot sun can be tiring. These miners take a break in a front-end loader bucket at a mine in New Mexico. (Murray)

coalfields until the current study was undertaken.

President's Commission on Coal
In 1977, the miners struck the Nation's mines for 111 days, the longest and most economically crippling strike since 1950.

In an effort to reduce the possibility of such major disruptions in the future, President Carter established the President's Commission on Coal by Executive Order 12062 on May 26, 1978. The Commission was directed to "conduct a comprehensive view of the state of the coal industry in the United States . . . " with a mandate to deal with several issues, including "matters pertaining to . . . health, safety and living conditions in the Nation's coal fields . . . "

Mindful of the report of the U.S. Coal Commission and of the Boone Report, the President's Commission on Coal decided to take a comparative look at the community and living conditions for miners. What changes had taken place since 1946, when Admiral Boone sent his Navy teams into the coalfields? To what extent were the miners still those poor and despairing people pictured in the Boone Report? How far—if at all—had the miners come in joining other industrial workers as members of middle-income and middle-class America?

Profile of the Work Force
American coal miners are a varied group, consisting of men and women of all ethnic and racial backgrounds—white, black, Hispanic, Asian-Americans, and American Indians.

That variety—with the exception of women and Indians—has long been characteristic of American miners. Years ago, however, it was also characteristic of them to be poorly paid in comparison with the rest of the industrial work force, not as well educated, and generally not part of the contemporary middle class.

Many of those patterns are changing. Today, the work force is no longer entirely male: of the 1,281 miners responding to the Living Conditions Survey, 97 percent were male, and that high percentage is likely to drop steadily, if slowly, in coming years.

Miners answering the survey are largely married (71 percent),

and some of their spouses (24 percent) work. The miners' average income is $19,442 and 53 percent of them have completed at least high school.

Miners often come from mining families. Of all respondents, 56 percent had a miner-father, 37 percent a miner-grandfather, and 12 percent a miner-great-grandfather.

The mining work force is also younger than it was a generation ago, partly because a slump in production in the 1960s brought few new miners into the occupation. That resulted in the so-called "missing generation" of miners. Of respondents to the survey, fully 41 percent are 29 years old or younger.

Finally, miners seem to like mining — for all the dangers and uncertainties and plain hard work of the occupation. There seems to be a deep feeling of personal satisfaction in having teamed with other miners to complete the day's work and in overcoming the adversities of nature. Of all respondents to the survey, 64 percent say they would remain in coal mining and 66 percent say that they would decide again, given the chance, to be a coal miner.

Minorities in the Coalfields

Coal mining has long attracted workers from across a broad spectrum of American society. During the waves of immigration in the 19th and early 20th centuries, many workers came from England and Europe to work in the anthracite fields of Pennsylvania. These were Englishmen, Scots, and Welshmen, Germans and Czechs — often the sons of miners — who were drawn to the United States by promises of better wages, better working conditions, and a better life.

In 1922, the U.S. Coal Commission studied the 1920 Census data and determined that of the 525,152 workers in the bituminous coalfields, 59 percent were native-born white, 8 percent were native-born black, and 33 percent were born in other nations. The study found that most of the foreign-born workers were from Italy, Austria, Hungary, Poland, Russia, France, "Slovakland," England, Yugoslavia, and Germany. A miner alive today recalls the era when these men lived and worked together in Appalachian coal company camps; union meetings were conducted in three languages.

Many children and grandchildren of these miners still live in the coalfields, some as miners.

The lure of profit drew men of various backgrounds into management and ownership, as well. A listing of biographies of leading coal operators in 1918, *Coal Men of America,* included among those prominent in the Illinois

coal trade these names and places of birth: Allais (France), Clinch (Savannah, Ga.), Dillarou (DeLand, Ill.), Evans (North Wales), Groenier (Holland), Haskett (Kokomo, Ind.), Holversheid (Canada), Hull (Canon City, Colo.), Kelly (Canada), Koerber (Chicago, Ill.), and Scholz (Germany).

In 1947, the Boone Report found in the bituminous coalfields of southern Colorado and New Mexico, "whole communities of people of Spanish-American descent," as well as Indians, Japanese, and — in Rock Springs, Wyo. — persons native born in or descended from as many as 40 nations.

Black workers have found employment in the mines for many years, particularly in Alabama and some south-central coalfields in Kentucky and West Virginia. The Boone Report estimated in 1947, that 9 percent of the mining work force was black, and that these workers, "in their search for economic freedom, sought out or were brought into the coal fields." However, in 1947 — and apparently today — some coal States with small black populations have no black miners. In 1979, these States were mostly in the Rocky Mountains. According to one 1978 Government estimate, about 3 percent of those employed in the coal-mining industry are black.

Black miners have faced a number of manifestations of discrimination over the years — in housing for their families, in schools for their children, and in separate facilities (such as bathhouses) for themselves. But they have received more equal treatment from the union and from miners at work in the mine.

The Boone Report commented that black miners "belong to the same labor unions as the white, receive the same rates of pay, and are as respected for their skills as any other group." In his classic 1947 study, *Inside U.S.A.*, John Gunther noted:

In the Alabama coal mines the United Mine Workers accepted white and Negro members both, and this precedent was followed when iron and steel were organized in Birmingham.

Most miners live in small towns in rural America. Left: Madisonville, in the heart of the western Kentucky coalfields. Right: Storefront announcements in Fairview, Ill., a mining town. (Wathen)

There also has been traditionally "no discrimination underground," according to one observer experienced both in mining and management. Once down in the mine, the men work together as team members.

Above ground, it has been a different matter. Until recent years, especially after Federal law prohibited discrimination, black miners were systematically blocked from advancement, according to contemporary accounts and academic studies. Blacks often were relegated to the lower paying jobs, and (particularly in nonunion mines) were the last hired and first fired.

Today, the situation for black miners is somewhat more encouraging. Coal companies have undertaken affirmative action programs to hire and promote black miners; these programs have met with some — if mixed — success, and black employees can be found across a fairly wide range of managerial jobs in the coal industry. Coal companies contend that the major problem in finding new black miners is that many black families have migrated to the large urban centers and that few live in the coalfields.

Until the early 1970s, the mining work force was completely male. A few women had worked in the mines in World War II, and some held office jobs in coal companies. But males held the mining jobs underground and in surface mines. That began to change about 1973 when women started to seek the higher-paying jobs in the mines. The entrance of women and American Indians into the coal industry — both relatively recently — are the subjects of separate chapters in this book.

Major Issues in This Report

Coal miners know the dynamics of their industry as well as they know the dynamics, the stresses, and the strains of the tunnel or open pit in which they work. The former may generate uncertainty while the latter engenders fear and pride.

They know the booms and busts of coal operations as new mines open and old mines close. They know the sudden shutdown of a small mine as the market becomes glutted with coal.

The boom-bust cycle, the uncertainty, the danger of mining are ever present. But the days of a sense of powerlessness on the part of miners, and of economic deprivation for substantial numbers of their ranks, appear to be in the past.

The miner today is concerned about earning a steady income for his or her family. The new miner also is concerned about health and safety, and seems more convinced than ever before that accidents can be reduced in frequency and that Black Lung can be reduced or eradicated.

This report examines the total environment of miners and their families, with an emphasis on community and living conditions and how they have changed in the past 30 or so years. Miners' incomes, the working environment, health and safety regulations, and other factors bear on these community and living conditions, as do policies of government, financial institutions, and other entities.

In the last analysis, it is the miners themselves who appear to have exerted the greatest influence over their own lives. They have found the land and built housing; they have found the financing and built health clinics. Some conditions and services — roads and schools among them — are beyond the power of private citizens or groups to create. They are the responsibility of government.

Short of that, however, miners and their families show resilience in accommodating to the occasional hardships of the rural life that most of them lead, and they show initiative in exploiting and enjoying the benefits of that life.

Coal Country. Top left: Radio station sign in Price, Utah. (Corn) Bottom left: Eastern Kentucky. (Elliot) Center; Civic club signs in Colorado coal country. (Corn) Right: Miner Robert Davis greets his daughter, Mrs. Katherine Cook, and grandchildren after work in Man, W. Va. (Corn)

HOTEL
WEST ERN
NW
139439
NW
NW
38407

Housing

Coalfield housing is bound to the land and the industry. Coal miners, unlike their counterparts of a generation or two ago, find that they have some choice in their housing. Most prefer to buy single-family homes, but many of them live elsewhere.

Renovated company house, mobile home, town apartment, suburban ranch-style home, wood frame duplex, adobe hogan, cinderblock shelter, prefabricated row house — even pickup truck and tent — these are all miners' homes in 1979.

Housing is rarely a simple matter for miners, and many must contend with critical problems: land shortages in central Appalachia; costly water and sewer installation; lack of new housing developments around some western coal boomtowns; and the short supply and high cost of mortgage money for many miners who are prospective home buyers.

One family that sought housing is the Oswalts. When they were house-hunting, Dan and Anne Oswalt had many distinct options in their semirural region of Pennsylvania: new housing developments, older and more conventional houses, mobile home parks, and former coal company housing. Their combined salaries were around $30,000, from his work as a shuttle car driver in an underground mine and her job.

As is common among young miners in these parts, the Oswalts opted for an old company house because the price suited their savings and budget. They considered it their "starter" house and, given that it was little more than a shingled shell without plumbing or electricity, were willing to make their own improvements. Mr. Oswalt explains: "When we moved here, we thought by buying this house, we'd work on it a little bit at a time. But there wasn't anything in the house with something not wrong with it. We had to start from scratch."

The Oswalts spent more than $14,000 renovating and modernizing their two-story, frame home in Lucernemines, Pa. They added yellow aluminum siding; installed plumbing for a new bathroom; finished the living room, den, and upstairs bedrooms with wood paneling; and put a coal-burning furnace in the basement. The couple and their three children have lived in this house since 1976.

The pastoral hills of Pennsylvania are gentle, and single-family homes like the Oswalts' are easily situated on level lots routinely connected to water and sewer lines and large enough for a vegetable garden.

Homes in north-central and south-central Appalachia, in contrast, are balanced precariously on steep mountainsides. Little flat land is available and floods threaten valley housing. Basic water and sewer services frequently do not reach these mountainside homes.

Miners living in the Illinois Basin and Rocky Mountain regions do not lack land for their housing. Competition with other workers in the Midwest and population booms in the West, however, make housing expensive and, at times, scarce. Western miners also contend with housing shortages near new sur-

Overleaf: In traditional coal communities, such as Smithers, W.Va., housing is located close to mining and transportation facilities because of a scarcity of suitable land. (Murray)

Coal miners' housing varies with the terrain and often depends on the type and amount of land available. This mobile home park in Oak Creek, Colo., (top), and the former company town of McClure, Va., (bottom), are typical of coal-mining communities in those regions. (Corn, Murray)

face mines, which often leave them with little choice other than housing in mobile homes, modular units, prefabricated units, and other shelters.

The Oswalts live much like other industrial workers in northern Appalachia and parts of the Midwest. Yet, they are not representative of all miners. Coal miners living in rural coalfields in central Appalachia or on the desolate western plains have the same or more income and buying power as steelworkers, auto workers, or rubber workers, but they may not be able to live like them because of limited housing choices.

The Coal Commission survey teams found ample evidence that most coal miners live in much better housing in 1979 than they did a generation ago. Steadily employed coal miners now have the financial means to afford some of the best housing available in their communities. Therefore, the improved living conditions of coal miners do not necessarily reflect conditions for the general population.

In some coal-mining areas, especially in central Appalachia, many pockets of poverty and substandard living conditions persist. The Living Conditions Survey and this report deal only with employed coal miners, a relatively small percentage of the population.

Furthermore, in some areas of Appalachia, even the increased salary levels of miners do not enable them to obtain the same quality of housing available to coal miners in other regions or industrial workers in metropolitan areas. The ownership by large companies of much of the land that is suitable for housing, the shortage of capital available for construction and mortgages, and the hazards of periodic flooding in many communities remain serious barriers to the widespread availability of good housing.

What Distinguishes a Miner's Home?

Aside from the round aluminum dinner pail on the kitchen counter and the pile of sooty work clothes ready for the washer, the fittings and feelings of a miner's home are usually indistinguishable from those of a nonmining neighbor's home.

Miners' homes inside are apt to be furnished with the same appliances and amenities as other American homes. Indoor plumbing and running water are commonplace, even though only a generation ago these were luxuries in coalfield housing, where privies and hand-dug wells predominated. Electricity, even in isolated Appalachian hollows, has nearly always been available in coal-mining areas because mine operations depend on it.

The Oswalts' home is furnished with a combination of matching Mediterranean-style living room set and more casual, overstuffed cushions and area rugs in the den. Mr. Oswalt claims that he has not yet finished renovating the kitchen, but, within arm's reach of the circular kitchen table is an assortment of time-saving gadgets — blender, popcorn maker, dishwasher — plus formica counters and heavy-duty floor covering. Most miners own televisions, telephones, refrigerators, and radios. They decorate their homes with new furniture, wall hangings, and carpeting.

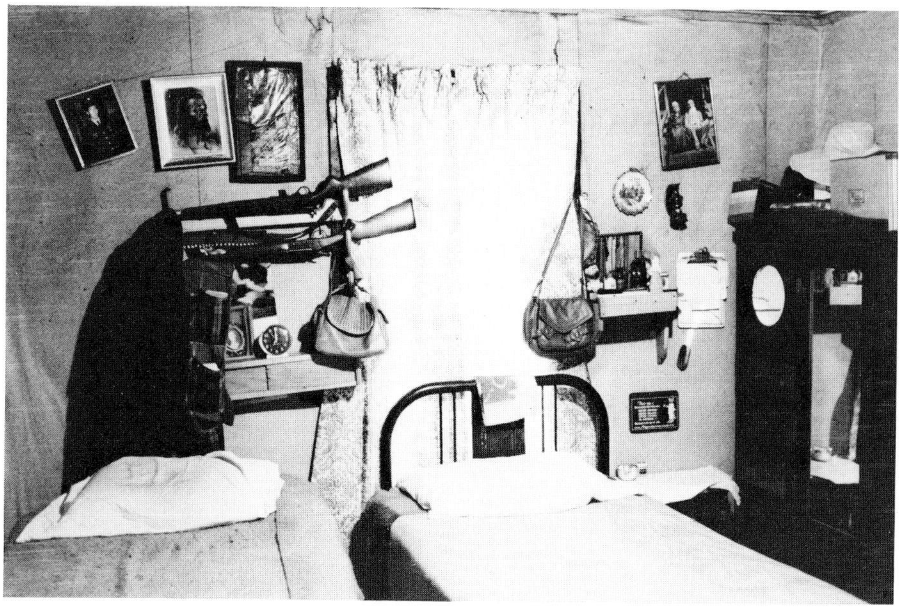

Left and center: The Oswalts' home, a renovated company house, is one kind of housing available in their part of southwestern Pennsylvania. Mobile homes, townhouses, suburban tract homes, and apartments are other options for northern Appalachian coal miners. The Oswalts spent $14,000 remodeling the home and equipped the kitchen with many small appliances, but they feel that the house still needs work. (Owre)

The interiors of the miners' homes are as different as their locations and give few clues as to the residents' occupation. Fine coal dust, always present in miners' homes of past generations, is now found only on dirty work clothes. These homes are in Gillette, Wyo., (top right), Hemphill, W.Va., (middle right), and Wheatfield, Pa. (bottom). (Murray, Corn, Simmons)

Living Conditions Survey Results

Coal miners are largely homeowners, according to the Living Conditions Survey. Seventy-four percent own their homes compared with the 1977 national average of 65 percent. Only 20 percent of coal miners rent their homes, compared with 35 percent of the population as a whole. Analysis of all miners responding to the survey showed that 68 percent live in single-family dwellings and 24 percent live in mobile homes. By comparison, 64 percent of all Americans live in single-family homes and 5 percent live in mobile homes.

Average monthly housing costs for all miners surveyed are $241. Average monthly housing costs for middle-income ($18,000 per year) urban Americans are $269.

Most miners responded in the questionnaire that they are pleased with their housing: 32 percent report they are "very satisfied"; 50 percent report they are "satisfied." Fourteen percent are "dissatisfied" and 5 percent are "very dissatisfied" with their housing.

They report that their water and sewer systems are generally adequate. Seventy-three percent report that they are either very satisfied or satisfied with their water system and 84 percent report they are very satisfied or satisfied with their sewer system.

Overall, 82 percent of the coal miners surveyed feel that their housing is better than or the same as that of most Americans. Seventy-eight percent say that their water system is better than or the same as that of most Americans, and 83 percent feel that their sewer system is better than or the same as that of most Americans.

Questionnaire results show overall satisfaction with housing conditions, but in-depth interviews with some miners who completed the questionnaire reveal another level of response. When interviewed about their housing conditions, these miners often described particular problems that they faced. These complaints are described in greater detail elsewhere in this report.

The most common complaints are about the lack of land suitable for building houses in Appalachia and the shortage of new homes in the western coal boomtowns. Miners in rural coalfields across the Nation contend that they must meet substantial downpayment requirements and pay high interest rates on mortgages. Both lending institutions and capital are in short supply, miners say, in their areas.

Regional Results

Despite problems with finding adequate housing in the West, a substantial majority of miners — 73 percent — say they are very satisfied or satisfied with their housing; that percentage is, however, the lowest of any region surveyed.

A bare majority — 51 percent — of miners in the West live in single-family houses. The rest live in mobile homes (37 percent), apartments (6 percent), and in an assortment of campers, recreational vehicles, hogans, tents, and motels. Average monthly housing costs for all western miners are $297, the highest of any region surveyed.

Appalachian miners responding in the Living Conditions Survey reveal that 71 percent own homes, 63 percent of them live in single-family houses, and 29 percent live in mobile homes. Compared with miners elsewhere, their average monthly housing costs are the lowest — $191.

Seventy-nine percent of the miners in north-central and south-central Appalachia report that they are very satisfied or satisfied with their housing.

Miners in northern Appalachia largely live in single-family houses, 71 percent. The same percentage own their homes. Nine percent are apartment dwellers and about 14 percent live in mobile homes. Their average monthly housing costs are $266.

Eighty-three percent of the northern Appalachian miners are very satisfied or satisfied with their housing.

Midwestern miners, compared with others around the country, comprise the largest percentage of home owners, 85 percent. They also comprise the most single-family home dwellers, 86 percent. Three percent live in apartments and 10 percent live in mobile homes. Average monthly housing costs are $244. Ninety-one percent of these miners are very satisfied with their housing, the highest percentage of any region.

Miners' housing, according the the Living Conditions Survey, consists mostly of single-family dwellings (68 percent), with mobile homes (24 percent) providing the primary alternative. Average monthly costs for homes such as these near mining areas around Craig, Colo., (left), Cokeburg, Pa., (top right), Jenkins, Ky., (center right), and Welch, W.Va., (bottom right) were reported to be $241 for all miners surveyed. (Corn, Wathen, Wathen, Corn)

Company Housing

They are time-worn reminders of the history of coal and of the miners and families who inhabited the isolated coalfields. These colonies of company housing, clustered in the valleys of Appalachia or perched on a windy plateau of the western plains, are recognizable by their relentless symmetry, each home mirroring its neighbors.

For some families, company housing created a well-defined, self-contained community that gave shelter and support; for others it was a lonely camp ruled by the company owner and the private army of coal and iron police that patrolled it. In either case, miners worked for the company, paid rent for the company house, went to the company doctor, and — using company scrip — bought food, clothing, and other necessities of life at the company store.

Company towns began springing up in the early 1900s, reached their heyday in the 1920s, and lasted until shortly after World War II.

Contemporary accounts of company housing paint widely different pictures; the quality and conditions of a miner's home depended largely on the individual coal company that built and maintained it.

One Appalachian company town in the 1940s was described as "Monotonous rows of houses and privies, all in the same faded hues, standing alongside the railroad tracks, close to a foul creek, or ... scattered houses on stilts, perilously perching, with their privies behind them, on steep hillsides."

Other company towns of that era offered better conditions. One writer described a community in southern Pennsylvania at the same time as consisting of:

... a double row of well-kept identical two-family, two-story houses alternating red and green, the sashes painted white,

with poplar trees, green lawns, and an alley running between the dahlias, cabbages, and neatly painted double privies of the fenced back gardens.

Coal camps date from the 1860s when, in the West, railroad companies imported gangs of immigrants to dig the fuel for steam engines and, in the East, coal was needed to drive the machines of the Civil War.

Wartime always spurred coal demand, and company towns multiplied between 1917 and 1920. There are no national statistics on exactly how many houses were built and rented by coal companies, however, and estimates mention only "thousands" of company houses.

In 1922, the U.S. Coal Commission reported that more than half of the country's approximately 664,000 coal miners lived in company housing. Much company housing was concentrated in north-central and south-central Appalachia, where up to 75 percent of the work force rented homes from employers. In parts of the Midwest, fewer than 10 percent occupied company housing. Today, few miners rent housing from coal companies; only one tenth of 1 percent of the miners responding to the Living Conditions Survey rent company houses.

Company towns, such as McRoberts, Ky., were built by coal companies to house employees close to mine sites and were usually in isolated rural areas. (Wathen)

Ca. 1946

Ca. 1946

Lease Arrangements

Lease and payment arrangements with miners were unique. Miners paid rent, which usually was less than what other industrial workers paid; charges for fuel, electricity, gas, and water, if available, also were small. The coal company customarily deducted rent and expenses from the miner's pay. These kinds of arrangements were typical as late as the 1940s.

The company leases recognized the importance of housing in the coalfields; a lease in 1946 stated:

> *. . . said houses under the control of said Lessor are maintained for the sole purpose of providing suitable and convenient homes for said employees engaged in the operation of said mines and other business of said Lessor; it being necessary for the persons employed by said Lessor to reside in said houses in order to be convenient to the work which they are required to perform . . .*

If the miner left the employ of the coal company voluntarily or by discharge, the miner and his family were required to vacate the premises immediately; if they did not, the company had the right, "without the necessity of giving notice and without liability for so doing, to remove said Lessee and his family and his and their goods and property from the premises."

Living Conditions

Company houses were usually built of wood frame set on foundation posts, with composition paper roofs, and clapboard or weatherboard walls. Some homes were little more than shells. According to the 1922 U.S. Coal Commission, only 14 percent had running water; about 3 percent had bathtubs, showers, or inside flush toilets; and approximately 66 percent had electricity or gas.

Twenty-four years later, the Boone study of 1,154 company houses found few improvements: only 10 percent had bathrooms

and fewer than half had piped water; 75 percent of the company towns surveyed used privies.

Coal camps offered more than regimented housing, however, and some former inhabitants remember a special spirit about them. In 1979, the wife of a Harlan, Ky., coal miner recalls growing up in a coal town:

I really liked it in the coal mining camps. It was just like one big family. All the houses were sort of close together. Everybody knew each other. If you had a problem, they had it. Everybody went to the same church. Everybody had the same occupation. We all had the same things in common. Nobody considered themselves better than others because they all made the same. I think everybody got along better.

Camaraderie aside, coal camp living was seldom easy, especially for housewives who constantly battled the dirt and dust thrown off by nearby mining operations. An account of life in the Pennsylvania coalfields describes the daily struggle:

Window curtains have to be washed continually; bureau drawers must have their contents covered; silver and brass tarnish overnight; and each spring ushers in an orgy of cleaning wallpaper, rugs, and upholstery. Even the daisies in the fields have dirty faces.

Top and bottom left: Houses in company towns were generally built at the same time and in identical rows. Some were squalid, drafty shacks made of cheap board and batten; others, usually the exception, were sturdy buildings set on concrete foundations and surrounded by lawns and gardens. (Lee)

Top right: Coal companies sold off much of their real estate after World War II and sometimes gave miner-tenants the first option to buy homes. A small number of company-owned homes still exist, however, such as this 11-room house in Kayford, W.Va., which a miner rents from his employer for $110 a month. (Murray)

Bottom right: The symmetrical identical architecture of former company housing still can be seen in older Appalachian communities. This street is in McClure, Va. (Murray)

Change Comes in the 1950s

Coal companies began to divest themselves of housing holdings in the 1950s as coal demand and production declined, miners left the coalfields, and taxes and maintenance costs turned housing into an unprofitable sideline.

Resident miners usually were given first option to buy, often at a nominal price. Some companies razed their towns. A coal miner in southern Virginia describes what happened to a wornout coal town near St. Charles:

There used to be another house sitting right in front of this one. Bonny Blue Mines, they had all these camp houses. And when they shut down after a period of years, they sold the camp houses for about $100 apiece and tore them down and built houses out of them. The lumber was the best lumber you could imagine. It all came from out West, you know, real good lumber and my father and brothers took three of them and used the best lumber and they built this house right here, and we're living in that house.

Few Company Houses Survive Today

Company-built houses survive today, although they are mostly privately owned. Of the 1,281 respondents in the Living Conditions Survey, only 7 rent their homes from coal companies. Many miners have purchased former company housing in the isolated communities of architecturally identical homes, personalized now by colors, additions, and improvements. The new owners build closets, install plumbing and insulation, finish basements, add bathrooms, and close in porches.

Telltale statistics on utilities and services, which documented the inadequacy of much company housing in earlier years, vanish with modern miner housing. Miners, including those who live in converted company housing, reported in the Living Conditions Survey that not only are water, electricity, gas, and other utilities virtually universal, but also that 90 percent have refrigerators, 85 percent have clothes washers, 81 percent clothes driers, 35 percent dishwashers, 81 percent stereos, and 93 percent televisions.

In some western coalfields, history is being repeated, but with a new twist. Companies, again, are building towns and homes to attract miners and families, but without the stranglehold leases, the monopolistic company store, or the private police force that characterized the past.

Wright, Wyo., for example, is a new town built by Atlantic Richfield Company for miners at the Black Thunder Mine, a big (20,000 tons per day) surface mine with a life expectancy of 40 years. Black Thunder is scheduled to reach full production in 1985, and local housing is in demand. In 1979, approximately 250 families lived in Wright in their own mobile homes, in company-owned mobile homes, and in private homes on sites prepared by Atlantic Richfield, built by private developers, and purchased by individuals.

Left: Company houses built in the 1920s often lacked plumbing, insulation, and a heating system. Miners who bought them decades later frequently have done extensive renovation, inside and out. This photograph shows a former company home in Vestaburg, Pa., before and after remodeling. (Wathen)

Top and bottom right: Wright, Wyo., is a new kind of company town. Atlantic Richfield Company bought and developed the land for homes for its miners at Black Thunder Mine, then sold the sites to a private developer. Miners (and others) in Wright may own their homes or rent mobile homes from ARCO. (Utrup, Murray)

Miners' Housing Varies by Region

Pocketed around the country—on the wooded hillsides and in the hollows of West Virginia, the rich farmlands of Pennsylvania, the thick pine forests of Alabama, the high plateaus of Wyoming, the grassy valleys between snowcapped mountains of Washington State—coal miners' homes vary as much as the surroundings.

Most miners share a characteristic of their trade that has been common since the days of the first company coal camps: housing is tied to the workplace. The automobile has loosened this tie, but the fact remains that miners must live within a reasonable radius of their mine.

That most often means life in rural America, a life that for many miners offers deep personal satisfactions, but for some brings hardships.

The West

The young coal miners come west, to the boomtowns of Price, Gillette, Colstrip, and others, to the "energy towns," for the \$20,000-to-\$30,000-a-year wages and a chance for the Good Life. Alone or with spouses and children in tow, they seek shelter; many settle at first for small mobile homes, new double-wide trailers, prefabs, modular units, and other variations on the theme of quick, practical housing. Later, the mining family may find a permanent home in town, build their own home, or buy a ranch on the open prairie.

Miners have set up house in the mountains of Colorado and across the plains of Wyoming since the 19th century, but their tenancy has often been short-lived as the changing demand for coal and the cost of transporting it forced them to seek work elsewhere.

Some housing in the West is makeshift, created by patching up wind-weakened mobile homes, adding coal-burning stoves to uninsulated prefabricated houses, or creating more space by knocking out walls and adding rooms. In recent years, the boom-bust cycle of the coalfields created colonies of instant shelters that were quick and easy to assemble, but were not sturdy or permanent enough to house the next wave of workers.

At first, builders and developers showed little interest in the opportunities in the booming coalfields. More recently, however, this situation has begun to change. Developers now sense that mines are operating over the longer term and that new housing is profitable as the influx of miners continues. Solid new housing, some of it custom-built in the style of contemporary western architecture, is rapidly appearing in the coalfields.

In the Living Conditions Survey, 74 percent of western miners say that they own their own homes. Housing by type shows that 51 percent live in single-family homes, while 37 percent live in mobile homes and the remainder in some other type.

Mobile home parks are scattered throughout the western coalfields. Rows of rectangular mobile homes in light shades of brown, yellow, and green aluminum siding line the outskirts and pack the residential areas of many developing coal communities. The units are close together and loud noises carry easily through thin walls.

Commonly, mobile home parks are too new, or the residents too transient, to cultivate the trappings of a neighborhood. There are few lawns, gardens, and play yards. A miner living in a mobile home park in Colorado describes his surroundings: "I've had people move in on both sides of me and I don't even have enough room to go out and spit without hitting the trailer next to me."

Many families buy mobile homes already furnished. Sofas, chairs, appliances, dinette set, carpeting, and lamps are standard features that give a cozy feeling. Kitchens are modern and efficiently designed with everything in reach. Counters, cupboards, refrigerators, and stoves are all within a pivot.

Top: Miners in the West often encounter housing difficulties because construction has not kept pace with population growth in the energy boomtowns. Prefabricated housing, shown here in Rock Springs, Wyo., is one way builders are trying to fill the housing gaps. (Corn)

Bottom: Mobile homes provide housing for 37 percent of western miners, according to the Living Conditions Survey. Trailer sales are often directed at incoming coal miners, as at this lot in Hazen, N.D. (Murray)

187A
189
188
187
186
179
176
175
280
279
273
272

Top and bottom left: Mobile home parks such as this one in Rock Springs, Wyo., dot the western plains, affording little privacy or space for landscaping. (Corn) Bottom right: George Coleman, who works at a surface mine at Kayenta, Ariz. (Murray)

Colstrip, Montana

The Sopher family is representative of western coal mining families who, because of lack of money, choices, or commitment to the area, first live in a mobile home and then move into a more permanent home.

Moving from the coal mines of Illinois, Ed Sopher, his wife Mary, and their two children came to Colstrip about 10 years ago when only 30 families lived there. They first rented a house, then moved into a mobile home. "You can't imagine — ask my wife — how much I hated that trailer house we were living in," Mr. Sopher relates. "All you could say for it was that it was a place to stay." Another Montana miner who has moved from mobile home to house explains: "Most of the newer miners live in trailers and have to go through a certain length of time before they can get into a house."

Ed Sopher describes the family triumph:

We just moved into a brand new house in Colstrip after living in Snyder's Mobile Home Court for 9 years. Are we ever happy. You can't begin to imagine what a difference it makes, having all this room, not being cramped up.

The Sophers now live in a prefabricated home with a three-bedroom, living-room-kitchen layout and all-weather siding with wood trim. They are finishing the house themselves by installing insulation, partitions, and walls. "We've had to learn wiring, plumbing, drywall to finish off the basement," Mr. Sopher relates. "We're going to make another bedroom, bathroom, and family room down here with the work area. You know, you learn how to do it yourself around this country. Why, you can't afford to do it unless you do it yourself — the price would be too high."

Gillette, Wyoming

In a population with an average age of 24, Joe and Donna Hoehs, 30 and 28 respectively, say they are the "older generation." They came here 3 years ago from Missouri, where he worked with his father on the family farm and she grew up as the daughter of a military man. Life there was established: homes had hedges and flower gardens, streets and sidewalks were shaded by trees, rolling farmland and hills began at the edge of town.

Gillette is different. Here is a prairie town situated on the yellow plains of Wyoming, where horses and cattle graze on lots next to mobile homes. The main street — Gillette Avenue — features western-style facades and neon signs side-by-side, and the noise of coal trains drowns out the car radios of cruising teenagers.

The Hoehs "barely" found an

apartment, Mr. Hoehs relates, and paid $347 a month while they house-hunted. They were struck with the intense boomtown atmosphere; transient construction workers were a high-spirited lot, and the Hoehs even witnessed their first fistfight.

The construction crews left and Gillette became home to the oil drillers, who had been there since the early 1960s, and to the newcomer coal miners, business people, and service providers. Now the Hoehs, with their own house, are established. Gillette feels to them like many mid-American towns, with large grocery stores, a regular group of friends, and a routine designed mainly to provide attention to their daughter, Merrit, born early in 1979.

Their living room is an eclectic assemblage of styles: rock wall veneer covering, ornate Italian-style chandelier, velour couch, dark wood coffee table with matching end tables, a swing set for the baby. Mr. Hoehs shrugs his shoulders over the living room, largely the creation of former owners. "This is a combination of every style in the world," he observes. On the book shelves are the *Bible* and a collection of great books. Television guides, the *Reader's Digest,* and James Michener's *Centennial* are current reading in sight.

Joe Hoehs works at the Carter Mining Co. Rawhide Mine, a vast new surface mine—2.6 million tons produced in 1978—with a life expectancy of 40 years. Donna Hoehs worked as a bookkeeper at a local bank until Merrit arrived; she misses the income and the social life of regular work, but now prefers to spend her days walking with the baby or working in the garden. Evenings and weekends, the Hoehs visit friends, play cards, attend company-sponsored dances, or explore new parts of the endless Wyoming landscape. Sometimes they talk of returning, eventually, to Missouri, to be closer to their families. For now, they regard Gillette as their home.

Left: Miners in the West often live near reminders of their work. Behind this neighborhood in Gillette, Wyo., is a coal tipple. (Murray)

Right: Donna and Joe Hoehs lived in a $347-a-month apartment when they first moved to Gillette, where he works at the Rawhide Mine. They now own a frame house in a quiet residential area within walking distance of the town center. (Berger)

Craig, Colorado

Jesse and Emma Mascarenas live in the western tradition. He is an outside utility worker on the day shift at the Empire Energy Mine. Wearing blue jeans and a blue-and-red checkered cowboy shirt, Mr. Mascarenas describes how he built the house from the foundation up; it took 7 years, and he is still finishing the porch. Their neighborhood has dirt streets and no street signs and ends abruptly where prairie begins.

In the living room, as wide as the house, Jesse's trophies for bowling and calf roping rest on the mantle and in a bookcase; there are dozens of family photographs — children, graduates, brides — as well as a fern hanging in the corner, furniture and fabrics in an orange motif, and a large tapestry of the Last Supper. A friendly Siamese cat springs to a visitor's lap.

East of town lives the Kamzalow family; they moved here from Golden, Colo., where they sold their home for $23,000. They bought a new house in Craig for $46,000 in 1976; it is a three-bedroom home on one-third of an acre. It has a steep driveway — "a real ski slope," Mr. Kamzalow says — and city water and sewer services. Because Mr. Kamzalow's salary has increased since he bought the house, he is not strained to meet the payments. With his wife, Pam, and daughters Dana, 5, and Valerie, 4, Mr. Kamzalow intends to remain in Craig, which he finds to be a pleasant place to live, work, and raise a family.

Central Appalachia

Ben and Katherine Cook feel lucky. Some young coal miners in Appalachia must be content with rented homes or mobile homes. The Cooks own their home, a yellow, wood frame house in Man, W. Va. Katherine Cook describes the home:

Everybody around here puts a great deal of time and effort and money into their homes. It's something we all do. We paid $7,000 for our house, which was a relatively small amount 2 years ago. We have two bedrooms downstairs and a combination playroom-guest room upstairs. This has all been added. When we bought this, it was a plain, five-room house with a bathroom. That was all. We have invested $15,000.

Katherine Cook says of a neighbor's house: "Next door

was an old brown house virtually dilapidated. The porch was falling in. It was terrible. Then a couple bought the place 2 years ago, and they have completely redone the whole house — all but the outside."

Although many homes in Appalachia appear outwardly decrepit, especially those constructed during the bituminous coal booms in the 1920s and 1940s, they may be extensively remodeled inside.

The Cook's home has been refurbished inside and out. The full-length porch is painted deep brown and faces a neatly mowed front lawn surrounded by a white picket fence. In the backyard, the Cooks have arranged a swingset for their two daughters, Kathy Rose and Tina Rae. The home sits along treacherous Buffalo Creek, which devastated the valley in 1972 when a slag dam at a coal mine burst and water flooded 16 towns, killing 124 persons. The Cooks moved here in 1977.

Inside, the Cooks have added wall-to-wall red shag carpeting and wood paneling to the living and dining rooms. Their kitchen is conveniently fitted with an array of appliances and decorated in lively print contact paper. The girls' room, furnished with white-canopied beds, is bright pink and green.

Left: Jess Mascarenas, a surface miner in Craig, Colo., spent 7 years building his home from the foundation up and has only to finish the porch. (Berger)

Top and bottom: Ben Cook, in front of his home in Man, W.Va., and with his wife and daughter, paid $7,000 for his house and has spent much more remodeling and renovating it. His additions include wall-to-wall carpeting and wood paneling. (Corn)

McRoberts, Kentucky
Gary Lamb, his wife Denise, and their 16-month-old son Gary Jr., live in Upper Tom Biggs Hollow in McRoberts, where old company homes have gone to ruin and been abandoned. Mr. Lamb describes his constant repairs:

You get one of those old houses, man, you got to remodel it inside and out. I tore the porch roof off. I'm getting ready to put some windows in. You're talking about thousands of dollars. We put in walls, floors, the ceiling.

The Lambs' row house was part of a community when McRoberts was an active coal-mining town. Today, it sits on a lot empty of shrubbery or landscaping, although dense greenery surrounds the community.

From the outside, the Lambs' home gives little indication of what lies in store for the visitor. A 2-foot-wide gash where a porch should be runs along the entire front side, translucent plastic covers two windows, and a bare concrete slab juts from the front door. But it has been meticulously remodeled inside. Hardwood floors, wood beams in ceilings and walls, paneling, and simulated brick add warmth and personality to the home.

Constant repairs — the "fixing, patching and making do" — are only part of the housing struggle. The Lambs also live with an antiquated water and sewer system. "The water's terrible. I boil mine, and I'll cook with the water and stuff, but we'd rather buy spring water to drink. Sometimes it'll run dirty for hours. You can get a glass of that water and it's got sediment in it." Mr. Lamb adds about the sewage system, "Some of these houses down this line, they don't even have septic tanks. They put their sewage right in the creek."

The Lambs have considered moving locally but not from the area, because Mr. Lamb is a third-generation coal miner and his grandfather and father live nearby. Mrs. Lamb summarizes their feelings about their home: "Gary is still thinking that he might have made a better deal if he had knocked the house down and dozed off a larger flat spot and started over again from scratch."

Mobile Homes in Appalachia
Families who cannot find homes in Appalachia often turn to mobile homes. The Living Conditions Survey reveals that almost 30 percent of Appalachia's miners live in mobile homes.

Brenda and Gillard Noonchester occupy a silver, green, and gray mobile home in Coeburn, Va., on the site of a former coal town, Crane's Nest, which was demolished in the 1950s.

Coal-mining families furnish their mobile homes with televisions, modern kitchens, furniture sets, and other fixtures, but they consider these shelters temporary. Brenda Noonchester talks about trailer life:

Well, it's easy to keep clean. You know, things like that. It's compact and everything, but I'd really like to have a new home. A lot more room than this thing. We've only got two bedrooms and we have a boy and a girl and we need another bedroom bad. We paid $15,000 for the trailer and the lot, about three-quarters of an acre. The trailer was second-hand. New ones are about $15,000 without a lot and we couldn't afford it. Everything comes in them, though. When we bought this, it was bare, there was no furniture at all. It's convenient to everything, but I'd much rather have a house.

The Noonchesters, by living in a rural area, have escaped the annoyance of crowded mobile home parks. "We lived in a trailer park before. You could hear the people in the next trailer arguing and everything. We're better off here by ourselves."

Regardless of what kind of home they occupy, Appalachian miners generally collect around small, rural communities where families may have lived for generations, and brother or parents may live next door.

Top and bottom left: Miners, such as Gary Lamb in McRoberts, Ky., modernize former company houses themselves. "You get one of those old houses," he explains, "man, you got to remodel it inside and out. " (Wathen)

Top and bottom right: Mobile homes shelter almost 30 percent of the Appalachian miners interviewed in the Living Conditions Survey. While residents complain about flimsy construction, poor insulation, and cramped spaces, they also say that their homes are easy to keep clean. The trailer park here is in Jenkins, Ky., (Wathen) Below is Kenny Jewel, a miner in Harlan, Ky. (Corn)

Northern Appalachia

Northern Appalachia — parts of southwestern Pennsylvania, eastern Ohio, and northern West Virginia — shares only its name with the more southerly mining regions. Here the homes are spread over rolling farmlands and mildly sloping hills instead of being wedged into narrow hollows or balanced on mountainsides. Miners live in renovated company housing, new suburban developments, apartments, farmhouses, and ranch-style bungalows. Some buy undeveloped land and build their own homes, often of their own design.

The Frank family — Dick, Marilyn, and their six children — live in a two-story, three-bedroom, wood-and-shingle house on a hill overlooking Hastings, Pa. Hastings is a one-traffic-light town bordered by potato fields and grazing land, and many mining families maintain their homes for generations. A screened porch where the children leave bicycles runs along the front of the house; in back is a large yard where the second family car is parked and the St. Bernard runs.

Although larger homes are for sale in the area, the Franks are staying put. Marilyn Frank explains why:

I would like a much bigger home. I remember when my husband went into the mines, thinking maybe now we can build a home. But with this many children, just the cost of utilities [is high], and if you want them to go to college and send them to parochial schools, you've got to sacrifice something, so we're sacrificing where we're living.

A few miners rent apartments, particularly when they cannot find housing they can afford to buy. The Living Conditions Survey shows that slightly fewer than 10 percent of the Northern Appalachian miners live in rented homes.

John and Dawn Basil rent a second-floor apartment in a house in Bridgeport, Ohio, situated on a main route used by rumbling, noisy coal trucks. Mr. Basil is a coal miner. This is their first home together — both are 18 years old — and their furniture is an eclectic collection of new pieces for 11-month-old Jessica and a second-hand sofa and chairs for themselves.

Although they would prefer their own house, the Basils chose an apartment because the $230 monthly rent is within their budget and Mr. Basil's plumbing repairs and handiwork around the place reduce their rent payment.

Bridgeport is not exclusively a coal town; it lies just across the Ohio River from Wheeling, W. Va., and is the home of workers from a wide variety of occupations. A surface miner who lives in Zanesville, Ohio, a city of more than 37,000 population, describes his neighborhood:

I don't feel like I live in a coal-mining community. My neighbors work as an auto dealer, another is a salesman, another's a mechanic, oil well driller, supervisor in a factory — lots of different jobs.

Left: Availability of housing for miners is less critical in northern Appalachia because there is more suitable flat land available than in central and southern Appalachia. (Simmons)

Top and bottom right: Dawn and John Basil of Bridgeport, Ohio, are one of the few mining families living in an apartment. With their daughter Jessica, they occupy the second floor of an old frame house along a major coal-hauling route. (Wathen)

The Midwest

"My dream was to get this house," declares Jeannie Wimsatt who with her husband, Bob, lives in the suburbs of Evansville, Ind. Their home, 15 miles from the Amax Ayrshire surface mine where Mr. Wimsatt works the midnight shift as a bulldozer operator, fits comfortably into a neighborhood of paved driveways, two-car garages, mowed lawns, and landscaped yards.

The Wimsatts moved here about 4 years ago, about the time that Mr. Wimsatt left factory work and became a coal miner. "We lived in a trailer for a while, then when Bob was in the Army we lived on the base in Fort Benning, Georgia, and that wasn't too nice. I've finally got my house, and I'm satisfied," says Jeannie Wimsatt.

The Wimsatt home is two-story, of brick construction, with bedrooms for themselves and their 9-year-old daughter, Angie. It is furnished in contemporary beige colors with distinctive, more personal additions such as an Early American dining room table and a 2-foot macramé elephant handcrafted by Mr. Wimsatt.

Many coal miners in the Midwest, like the Wimsatts, own their homes and are pleased with housing options. The Living Conditions Survey shows that more Midwest miners live in houses — as opposed to apartments or mobile homes — than miners in other regions, and that slightly more than 70 percent of those surveyed own their homes.

Their homes are not likely to be former company houses. In 1922, when many coal towns were already established, the U.S. Coal Commission reported that fewer than 9 percent of company houses around the country were in the Midwest.

While other coal miners were living in company "patches," midwestern miners were settled in farming communities, living in homes with large vegetable gardens and surrounding acreage that stayed with a family through generations.

Lewiston, Illinois

Miners here blend with the farming community. Ken and Linda Howerter, a coal mining family for 17 years, live outside Lewiston, Ill., amid grazing land and barn country. Their two-story, white frame farmhouse is more than 80 years old and sits on a full-acre lot dotted with thick, shady trees that was once part of a 100-acre dairy farm. They share their four-bedroom home with their four children and Mr. Howerter's father, who has also mined and farmed this region.

The Howerters' furnishings fit their rural surroundings. A porcelain pitcher and basin on a night stand, an antique wall mirror, a steel frame bed, and well-used upholstered chairs all speak of comfortable country living.

Left: Bob Wimsatt lives in a single-family home in the suburbs of Evansville, Ind. More miners in the Midwest (more than 70 percent) live in single-family dwellings than do miners in other parts of the country. (Simmons)

Right: The Howerter home in Lewiston, Ill., is more than 80 years old and once was part of a 100-acre dairy farm. Many Midwest miners, like Ken Howerter, tend farms evenings and weekends. (Wathan)

Major Issues in Coalfield Housing

Beyond the shadows of the urban apartment buildings and the high-rise condominiums, tens of millions of American families cope daily with the problems of maintaining a home.

Few houses are trouble free, few living places entirely benign. Most families face high site and building costs, and some face poor construction and insulation of the house or inadequate water or sewer services.

Many families also encounter the dilemma of not only wanting a roomier house as the family enlarges with children or aging parents, but also wanting to save money for educating the children.

Coal miners are not exempt from these concerns; they may even experience an added measure of anxiety over finances because of the economic ups and downs of the coal industry.

Miners also share with millions of other Americans the pleasures and inconveniences of rural life. Life in the coalfields is life away from the noise and pollution of the cities, away from crowds, closer to nature, where people pride themselves on their independence, and still make things with their hands. Rural life may also include such factors as shortages of expert carpenters and plumbers, and high costs for the installation of telephone or electric lines.

In the coalfields of the West, water service is limited and new housing is at a premium. In some eastern coalfields, land suitable for housing is not readily available, and developers shun the low profits of rural projects. In many rural areas, capital is in short supply in lending institutions, and miners must turn to commercial banks, which may require downpayments of up to 30 percent for a mortgage extending over only 10 years. For most miners, such terms are

onerous. In some places, water and sewer services are of uneven quality or are nonexistent. Everywhere, costs are inflated.

Despite these problems, 80 percent of the miners responding to the Living Conditions Survey say that they are "satisfied" or "very satisfied" with their housing. For miners who live in single-family homes — 68 percent of the sample — the satisfaction rate is 87 percent.

Miners who are not satisfied with their housing, however, may face formidable hurdles in finding good housing.

Homes in Appalachia are close together and seemingly stacked on hillsides such as this one in Welch, W.Va. Rugged terrain and ownership of large tracts by coal and land companies leave little land for housing above the flood plain. (Corn)

Housing Problems in the West

A look at two coal boomtowns in the West provides a picture of the problems in that area: plenty of land, but too few houses built; plenty of income, but too little saved. When the boom cycle is in full swing, demand quickly outpaces supply.

Gillette began as a railroad town in the 1890s; until the 1960s, agriculture, railroading, and coal mining supported the community. An oil boom began in the early 1960s, peaking about 1969; the population increased from 3,500 to 7,000 in those years. Then, in late 1973, the coal boom hit. In one brief period, from February to December 1977, the number of households in Campbell County increased 25 percent. The population in 1979 was about 13,000 within city limits with another 4,000 in immediately adjacent areas of Campbell County.

Campbell County is mineral rich, with an estimated 4 percent of the Nation's coal. The tax assessment base of $596 million provides school, hospital, and recreation facilities and county roads. The city provides water and sewer services and city streets.

Mobile homes are everywhere in Gillette: in established residential areas, next to businesses in town, and in new subdivisions outside town. People generally rent mobile home lots, which go for about $100 per month. Most mobile home parks have available space and mobile home sales offices stretch down the main highway. There is no shortage of this kind of housing in Gillette.

In Campbell County as a whole, 53 percent of the families reside in mobile homes. (In the West as a whole, coal miners responding to the Living Conditions Survey report that 36 percent live in mobile homes.) In the Gillette planning area, as of the latest inventory in December 1977, of 5,295 dwelling units, 2,438 (46 percent) were mobile homes; 2,116 (40 percent) were single-family dwellings; and the remainder were duplexes, multifamily dwellings, or rural households.

The housing inventory showed that, for all dwelling units except mobile homes, 80 percent were in standard condition, 17 percent in deteriorating condition, and 3 percent deteriorated or dilapidated.

Land and capital are both available in Gillette. An estimated 2,250 unimproved acres around the city are zoned for homes. Mortgage terms in mid-September 1979 were at least 20

percent down and 11.5 to 12 percent interest plus 1½ points.

Land is plentiful around Craig, Colo., too. The coal boom hit Craig in the mid-1970s, when the captive mines were opened to feed the coal-fired utility plant just outside town. Coal mining has become a stable industry there.

"People saw the boom cycle with coal production," says K. Kyle Revelle, President of the First National Bank in Craig. "Then everybody waited for the bust. Large developers held back until just last year. Now it's catchup time and they're 2 years behind." In 1979, 70 percent of home sales were new houses.

Lots cost from \$12,000 to \$15,000. Prices range from \$15,000 on average for a mobile home 14 x 70 feet to \$55,000 on average for a three-bedroom house. With 14 percent of its real estate portfolio in mortgage loans, in September 1979, the bank offered home mortgage money at 11.5 percent to 12 percent plus 2 points for a term of up to 30 years, with 10 to 20 percent down. For mobile homes, the bank offered loans, usually for 10 years, at 14.5 percent with 20 percent down. These notes were in effect when the prime rate was 13.25 percent, before the severe actions of the Federal Reserve Board raised the prime rate in October 1979. Mobile home loans outnumbered home mortgages 2 to 1.

Top: Housing shortages in the West often leave miners with little choice of type of home. This neighborhood in Gillette, Wyo., shows permanent housing interspersed with mobile homes. (Murray)

Bottom: This mobile home park near Gillette has the requisite services but is still too new for sidewalks. (Murray)

Finding Suitable Land in Appalachia

For miners in north-central and south-central Appalachia, everything depends on obtaining land.

There's no land available — it's just not for sale. That's true for the seven counties. A fine piece of land to build on belongs to the coal companies. If they increase coal production, where would they put the coal miner? Put him on top of the mountains — that's the only place left here in Buchanan County.

Although this southern Virginia miner is talking about one area of Appalachia, the problem he describes exists throughout the eastern coalfields. Land suitable for housing is difficult to find.

The land shortage in Appalachia is, in part, attributable to coal companies, railroads, and other corporations owning much of the coal-rich acreage. With future plans to mine their holdings, companies prevent their unimproved properties from being developed. They are reluctant even to lease their real estate and risk creating ill will when they must later dispossess families for mining operations.

Statistics for land ownership are often buried in inaccessible, or untraceable, county records. However, spot studies of Appalachian counties reveal a pattern of large corporate land holdings:

- Forty-eight percent of Harlan County, Ky., is owned by corporations.
- Thirty-three percent of privately owned land in West Virginia is owned by 12 corporations.
- At least half the land in 27 counties in West Virginia is owned by corporations.
- Thirty-five percent of six coal-producing counties in southern Virginia is owned by corporations.

In addition to obtaining surface ownership, companies also buy or lease mineral rights to select tracts, thus further reducing the amount of land available for housing development. A 1974 journalistic study of land holdings in West Virginia estimated that, in addition to outright ownership, companies controlled leases on an additional 4 million acres and that corporate lease holdings would increase by 500,000 acres each year.

A miner in West Virginia explains another element in the Appalachian land and housing shortage:

There's just not that much land to build on. You get down too low and you're gonna get flooded, too high and you can't get to it in the wintertime. We've really got a problem.

Appalachia is hilly country. Steep, serrated mountains covered with thick brush and trees jut from flood-prone valleys and form narrow hollows. Highways, houses, railroads, buildings, and

businesses all compete for the same slivers of flat land. Land sloping at more than 15 degrees off the horizontal is too steep for safe home construction without using unusual and expensive techniques; an estimated 75 percent of the land in central Appalachia exceeds that slope. Some areas are impossible to reach. Only 2 percent of the land in Pike County, Ky., is level enough for building, and half that tends to flood.

Flooding threatens many Appalachian residents. Eighty percent of the land suitable for housing in eastern Kentucky is in the flood plain. The Tug Fork of the Big Sandy River, for instance, which runs down the Kentucky-West Virginia border, is notorious for flooding. The U.S. Army Corps of Engineers reports that the river valley is one of the most flood-prone areas in the country. Every 15 months or so, some section of the valley is flooded. When the river overflowed in April 1977, 1,372 families — including many

Flooding of what little flat land is available for housing is a constant danger for Appalachian miners. Residents of this trailer park near Pound, Va., must be ready to move quickly when the creek begins to rise. (Murray)

mining families — were left homeless.

High development and site preparation costs mean that few builders come into the area. Local entrepreneurs are unwilling to raise the capital and take the risk of home building.

Because of the financial and physical risks, Federal, State, and nonprofit agencies are building many of the new homes that are being built. The Human Economic Appalachian Development Corporation, located in Berea, Ky., is among the organizations that are coordinating acquisition of land, money, and other resources necessary for new housing developments.

Another is Coal Field Housing, Inc., a nonprofit organization seeking to improve housing for miners and other residents in West Virginia. In 1979, the organization was beginning a project to construct 67 homes near Whitesville, W. Va. The organization obtained land by donation from a company and made it available to builders. The new homes for miners and other residents were to cost between $38,000 and $47,000. The Federal Government, the State, the West Virginia Housing Development Fund, and an anonymous source were providing financial support.

West Virginia also is the site of an unprecedented effort to obtain land for housing by invoking the power of eminent domain. The State first sought to purchase land in Mingo County from a Philadelphia, Pa., development corporation to use for housing for families left homeless after the Tug Fork River flood of April 1977. When the company demanded a price that the State considered unreasonable, the State initiated proceedings to obtain 83 acres of land by eminent domain. The trial court ruled in the State's favor. (The development corporation, a railroad, and a steel corporation combined own more than half of Mingo County.)

Despite all the problems, coal miners remain fiercely rooted to their precarious mountainsides and unpredictable river valleys. Housing improvements must come to them. What little land they own is passed through generations like family heirlooms.

A young miner in southern Virginia describes his ties to the land:

I've often asked that question, why people live here. At times in the winter, when it gets so rough around here, you know, I say, "Why? Why am I stuck on the top of the side of this hill? Cause my ancestors were here." There's nothing keeping me from leaving except myself. There's a certain peace of mind that goes along with staying at a place like this for a long time.

Financial Issues

The high cost of housing hits every homeowner, but coal miners — particularly miners living in rural areas — encounter additional financial hurdles when buying a home.

Financial institutions in rural areas sometimes are reluctant to commit mortgage money. Although individual lenders set the amount of downpayment, length of time allowed for repayment, interest and penalty rates, studies show that mortgage terms are more stringent in rural than metropolitan areas. Miners experience this conservatism first-hand.

"Banks want a 30-percent downpayment if you want to buy a house. If you're going to buy a $60,000 house, that's a lot of money. Myself, I don't have enough to put 30 percent down. I've just got a little bit of what was left out of last payday," says a miner in Hillsboro, Ill.

A 1976 report on housing costs in Greene County, Pa., a region populated largely by coal miners, showed that mortgages required at least 20 percent down and ran for an average of 12 years. Mortgages in urban areas allow for smaller downpayments and 25-year terms.

A report on lending practices in rural and urban areas of Kentucky disclosed similar patterns. Rural lenders required an average 30 percent down; urban downpayments averaged 25 percent.

Left: In spite of housing difficulties, Appalachian miners cling fiercely to their isolated hollows and precarious mountainsides. Estel "Ratchett" Mullins built this house near Clintwood, Va. (Murray)

Right: Financing for housing in rural areas is often difficult to obtain because of lack of savings and loan associations. The Harlan Federal Savings and Loan Association is one of just nine in a seven-county area of eastern Kentucky coal country. (Corn)

Rural areas also suffer from a shortage of lending institutions; thus miners may have no access to capital or no choice of lender. Savings and loan associations (S&Ls) are the primary source of mortgage money throughout the country, and they are clustered in metropolitan areas. Greene County, Pa., for instance, has one S&L. Neighboring Fayette County, another coal-mining area, has no S&L. In seven Kentucky coal counties, only nine S&Ls exist and one county, Letcher, is without one. A survey of 147 coal-producing counties across the country shows that 91 have just one S&L or none at all.

Without savings and loans, miners have few other choices because urban institutions rarely lend outside their regions and rural commercial banks seldom make housing loans.

Some miners take advantage of the home financing arrangements available through the UMWA Credit Union. This organization, which is not formally affiliated with the union, has a small program of providing loans at 12 percent, with 20 percent down, for periods of 10 to 20 years (terms as of mid-September 1979). The Credit Union does not charge points or penalties. In September 1979, the Credit Union had about $3.5 million out in mortgages, with individual borrowing reaching as high as $50,000. The steady rise in the prime lending rate in 1979 apparently was responsible for an increase in miners' applications for mortgage loans, according to a Credit Union official.

Not all miners buy their homes. About 20 percent of the respondents in the Living Conditions Survey rent. Sixty-one percent of these miners say that they are very satisfied or satisfied with their housing. The largest group of renters — 26 percent — live in central Appalachia. Midwest miners who rent — 13 percent — are the smallest group.

Miners frequently mention prohibitively large downpayment requirements as the primary reason for renting. Some miners, however, prefer to rent. Willard Long in Kayford, W. Va., who lives in a fully carpeted, 11-room company-owned home, says:

With the setup I have here with the company house, I see no use in buying one. This way I can take my money and put it in savings. The cheapest house around would be about $20,000 for a four-room, five at the most. Owning a home right now is too great an expense. I prefer to rent. I only pay $110 a month.

Water and Sewer

Fifty years ago, water and sewer service in coal towns meant a central well serving multiple families and rows of outhouses next to a creek. These primitive services have been replaced by pipelines and modern bathrooms. Nevertheless, some older, more isolated coal communities still suffer from inadequate and deteriorating systems.

Miners responding to the Living Conditions Survey are generally content with water and sewer systems: 75 percent (water) and 85 percent (sewer) say they are satisfied or very satisfied. Unsatisfied miners are commonly faced with such water problems as low pressure, interrupted service, highly treated or polluted water, or the necessity of having to haul water from a distant source. Problems with sewer systems are backed up pipes or drainage of untreated waste into local waterways.

The hilly Appalachian countryside is ill-suited to standard pressure or gravity-flow water systems, and some miners never receive service. A Kentucky miner describes his community:

We've got people who live so high on the hill, they've never had water. They just hang up there and the water never gets to them. The city said it can't afford to put water there, so people have to haul water.

Sewer systems in some rural communities have just the opposite problem: they flow too readily into creeks and rivers. Miners and families throughout Appalachia frequently talk about public and private sewage and solid waste disposal systems that pollute streams.

Far left: In some Appalachian communities low water pressure, caused by antiquated and leaky pipes, can make fighting a fire more difficult. (Wathen)

Center: Solid waste creates eyesores in rural areas where there is no regular collection. (Wathen)

Right top and bottom: Privies have largely disappeared from the country's coalfields and have been replaced by modern bathrooms. Today, they serve mostly as reminders of the harsh conditions of the past. (Margolis, Murray)

New Legal Requirements

Inadequate water, sewer, and waste disposal systems not only cause pollution and health hazards but also they may hamper housing development in rural areas. Some communities are starting to require utilities to be in place before allowing homes to be built.

A city official in Logan, W. Va., describes how water and sewer service is changing: "People would get land from their parents to put a trailer or house on it, or they would move into a trailer park, and they end up with houses or trailers jammed together and then they wonder why the water and sewer never gets better." Referring to a new housing project, he says, "It's the first time someone has said that in order to have housing, you need utilities first. That's a revolutionary idea for this area."

Water in the West

Water has a special significance in the West where stretches of deserts and plateaus are so arid that water is a scarce and valuable commodity.

More than one-third of the Western miners responding to the Living Conditions Survey are dissatisfied with their water systems. They frequently complain about water polluted by sediment and livestock and about existing systems not being able to keep up with boomtown growth.

Coal boomtowns in the West often resort to rationing during dry spells to assure that water is used primarily for drinking and for public health and safety purposes. In the summer in Gillette, for example, outdoor watering of lawns and plant beds is limited to a designated 4-hour period every other day. Local residents there also complain that the quality of water is poor; the water is too hard and too high in sulfates to meet the aesthetic standards of the Environmental Protection Agency under the Safe Drinking Water Act.

The Principal Housing Alternative: Mobile Homes

Mobile homes have swept the coalfields. They cover hillsides in Wyoming in row upon row of rectangular, aluminum-sided boxes, and squeeze into the narrow valleys of Appalachia. The 1947 Federal survey of 2,028 miners' homes makes no mention of mobile homes. Today, about 24 percent of the miners responding to the Living Conditions Survey live in mobile homes, making these the second most common form of housing in coal communities.

Almost 32 percent of the miners answering the Living Conditions Survey do not live in single-family houses. Although some live in apartments, and condominiums and townhouses, their basically rural communities offer few alternatives other than single-family dwellings and mobile homes.

The largest group of miners living in alternative housing is in the West, where almost 49 percent cannot afford, cannot find, or do not want a single-family house. In boomtowns around Price, Utah, houses frequently are in short supply and miners must settle for mobile homes. In Sunnyside, Utah, mobile homes comprise almost 46 percent of all dwellings. The Living Conditions

Left: In Western communities, availability of water service is a key factor in planning housing developments. Water supplies are discussed at a city council meeting in Gillette, Wyo., where use of water is restricted during the summer. (Murray)

Right: Mobile home parks sprout up quickly near western surface mines. Many of them, like this park in Rock Springs, Wyo., lack landscaping trees or grass. (Corn)

Survey shows that 36 percent of the responding western miners live in mobile homes. In western coal communities with extreme housing shortages, some miners even pitch tents in nearby national forests or sleep in trucks or vans parked at the mine.

Although miners frequently complain about their mobile homes, about two-thirds of these residents report in the Living Conditions Survey that they are satisfied with their housing.

In parts of Appalachia, mobile homes are the only type of available housing. A realtor in Logan County, W. Va., remarks, "It's been a seller's market for a dozen or more years. I sell three times as many mobile homes through my agency than I do homes." His comments are borne out by figures. In Pike County, Ky., more than 90 percent of the housing additions between 1970 and 1978 were mobile homes. In Harlan County, Ky., mobile homes comprised 70 percent of the housing additions in the same period. The number of mobile homes in Lee County, Va., is estimated to have increased 350 percent between 1971 and 1978.

A mobile home is often a young miner's first home. According to the Living Conditions Survey, almost 60 percent of the miners in mobile homes are under 35 years old and half of this group is under 26 years old. An example of this group is Jerry Cline, 29, a coal miner, who with his wife Sue lived with his parents when they were first married, then moved to a rented mobile home. They are now firmly settled in their own 12 x 60-foot mobile home in Prosperity, W. Va., in the middle of an apple orchard.

The Price of Impermanence

The mobile home boom is attributable in large part to the shortage of other housing and to

However, on top of base price, mobile home residents also must pay rent for space and utility connections.

"You have to rent a piece of land for about $50 a month plus your trailer payment. A good one will cost $15,000, maybe more, and a doublewide will cost you even more than that," a Pike County, Ky., miner says.

A miner living in a mobile home park in eastern Utah says he pays $75 per month for space plus purchase payments of between $250 and $300. Across the border in Colorado, one miner had his pad fee raised by $35 a month. "You can't even find a place to park a trailer," he says. "You can buy a new trailer in Denver for less than the price of an old one here because of the cost of space."

the comparative low cost of a mobile home. A miner in central Pennsylvania explains:

Young people are moving outside town and buying trailers because they can't find a place to live. A lot of young miners buy trailers. It's an easy way to get started and not fool around with mortgages.

Western miners also settle into mobile homes because of cost. A surface miner who lives in Colstrip, Mont., tells why so many of his neighbors live in mobile homes:

Most of the newer miners live in trailers and have to go a certain length of time before they can get into a house. Really, you could say there's two classes of miners here—those who live in trailers and those like me who live in homes with kids and who have a supplemental income to get by.

Borrowing money is also easier for the mobile home buyer than for the single-family home buyer. A Kentucky miner says:

When you talk about a loan for a house, you really have to have $20,000 to $30,000 tied up in a savings account before they will even consider it. Or that much in collateral. But with a mobile home—anybody can get a mobile home.

Mobile homes cost less than single-family houses: a national average 1977 price of $14,000 versus $37,000 for a house.

Sam Spadero, an underground miner, and his wife Doris have lived in their mobile home in Washington, Pa., since they were married more than 2 years ago. (Wathen)

Miners Share Housing Concerns

Miners responding to the Living Conditions Survey generally are content with their total housing situation — the type, location, and size of their homes, the cost, and the availability and quality of utilities. Eighty-two percent report that they are "satisfied" or "very satisfied" with their housing situation.

Yet, indepth interviews with miners and their families disclose a different assessment. Although generally satisfied, miners also talk about housing shortages, about not being able to find, buy, or build the kinds of homes they prefer, about bad-tasting water, water rationing, or substandard sewer service, and about all the individual circumstances that make their homes less than ideal.

Some coal miners face critical housing problems. In central Appalachia, suitable and available land for building homes is scarce. Western miners living in coal development areas also encounter housing shortages because construction and rentals cannot keep pace with rapid population growth. All miners face substantial downpayment requirements and rising interest rates that may be too high for them.

Nevertheless, the days have vanished when coal miners and shanties, coal miners and dreary company houses, and coal miners and cold, cramped rooms were all considered synonymous. The homes of today's coal miners and their families, although not trouble free, firmly establish them in the American mainstream.

Alternatives to single-family homes and mobile homes for miners and their families encompass a multifamily dwelling in Hemphill, W.Va., (left), a converted box car in Oak Creek, Colo. (top right), and an apartment complex in Jasper, Ala., (bottom right) (Allen, Corn, Murray)

Top left, center and right: A miner and his home: Vivian and Marie Easton, Decota, W.Va. (Murray)

Bottom left: A miner and his home: Roy Longwell, Fairmont, W.Va. (Wathen)

Bottom right: A miner's home: Jerry and Carol Jandreau, Gillette, Wyo. (Murray)

Y
CAN'T B

Health And Safety

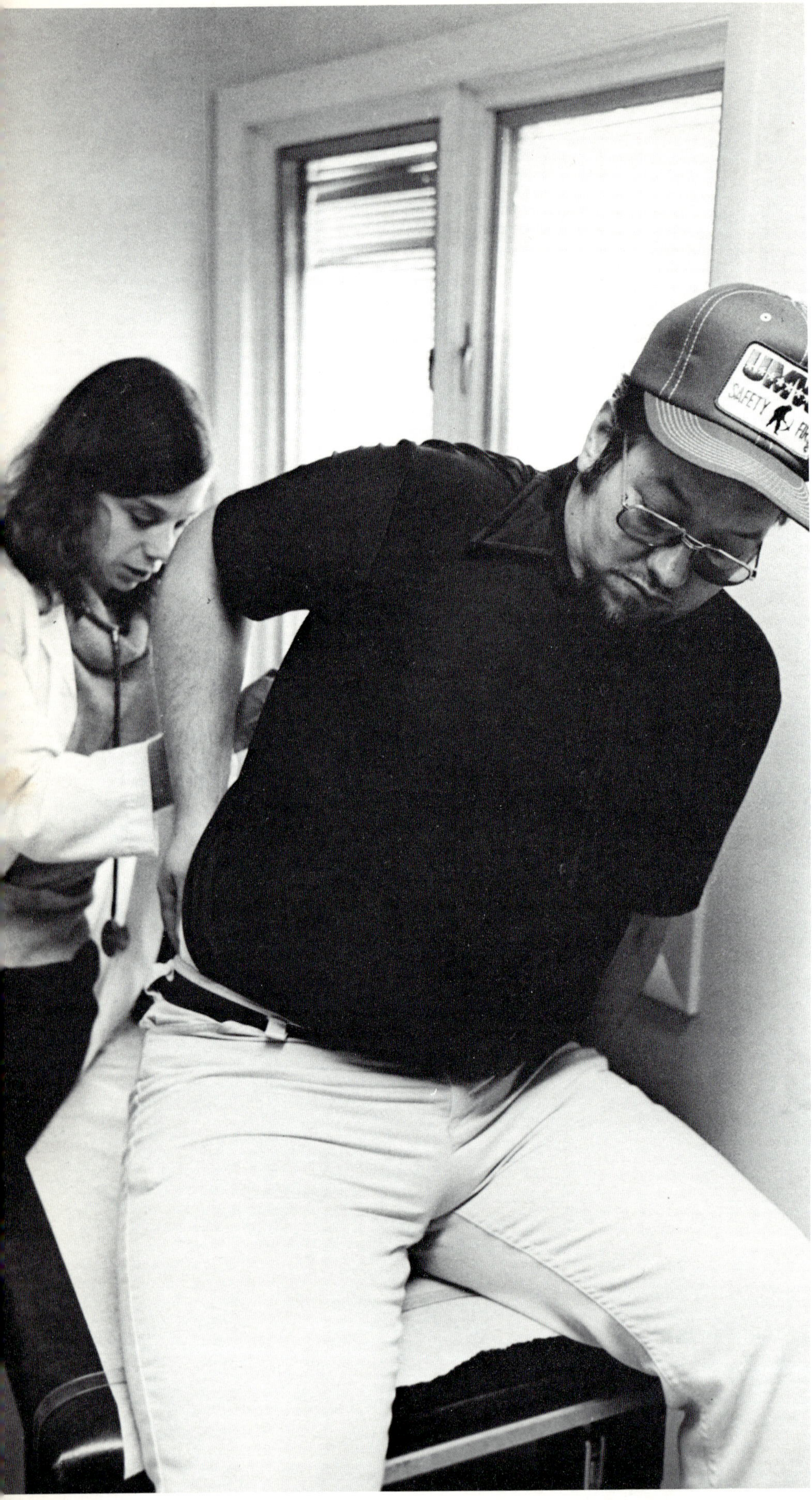

As the 1970s close, the Nation is engaged in a major debate on the issues that swirl around health care: the quality and quantity of physicians and other practitioners, the availability of the latest technology in medical sciences, and, above all, the cost.

American coal miners long were neglected in the area of health care for their families and industrial medicine for themselves. That neglect was so pronounced that it led to the massive Federal study of living and health conditions in the coalfields in 1946 — the Boone Report.

Conditions since the Boone Report have changed dramatically, largely because of the efforts of the miners and their union — the United Mine Workers of America (UMWA) — but also because of contributions by the Federal Government, States, and coal companies. The Federal Government and the States established a network of regulation aimed at ensuring health and safety in the mines; many coal companies, for their part, exhibit a new and enlightened attitude toward the health care requirements of the miners.

Miners, too, have changed. Miners live daily with physical danger and are aware of the importance of good medical care. But, observers note, miners had little conception in 1946 of the quality of health care that was beginning to be available to many other segments of post-World War II American society.

Thirty-two years later, coal miners have learned much about the intricacies of the health care delivery system in the United States. They have succeeded in obtaining for themselves a quality of health care comparable to that of many sectors of the industrial population.

Health care in the coal counties of the Nation today unquestionably is better than it was at the time of the Boone Report. Indeed, data analyzed in this

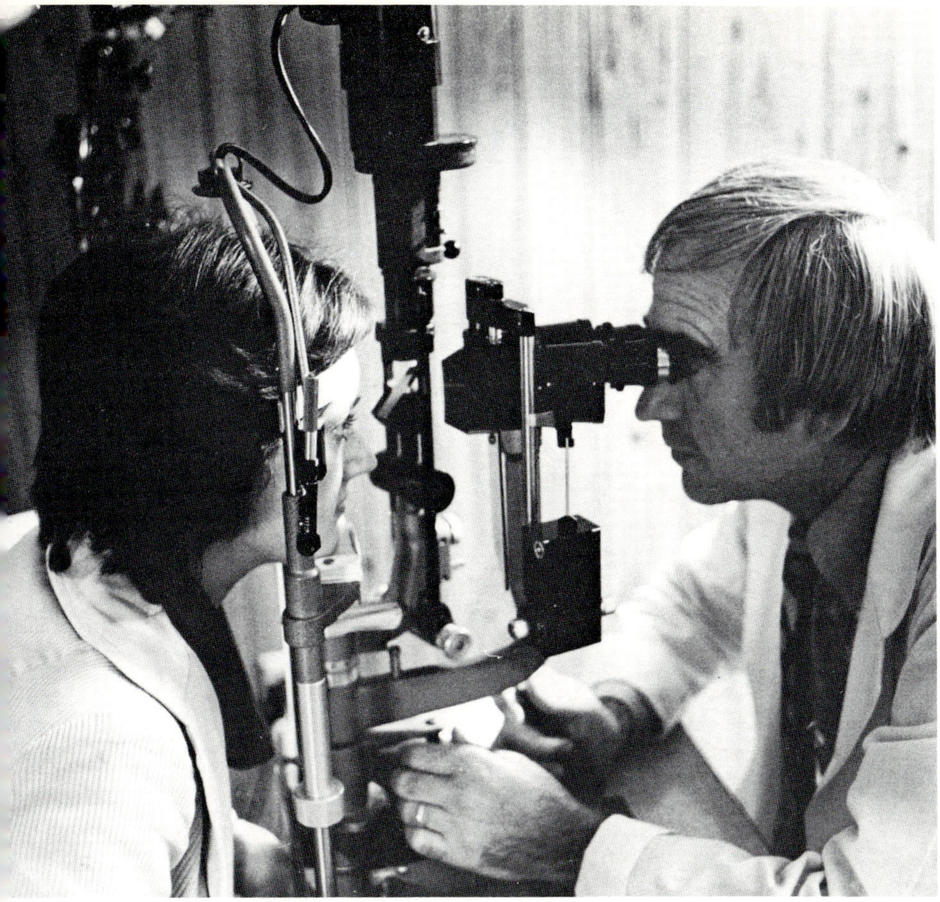

chapter indicate that, while medical care available to miners remains substantially below that available in large metropolitan centers, it often exceeds that available to residents of rural counties generally.

Lack of medical personnel and facilities and a shortage of specialists and specialized equipment are among the principal health care problems facing miners today.

Most miners are protected by health care plans financed by major insurance carriers. This gives miners financial access to the best care available in the region and the increased mobility of most miners allows them to travel for specialized care available outside the region. However, it should be kept in mind that these options may not always be open to the miner's nonmining neighbor.

Overleaf: When immigrant labor was cheap, some coal companies took the view that miners were easy to replace. Today, as this sign indicates, the health and safety of miners have become important concerns in the coalfields. This change has come about through the continuing efforts of the Federal Government, the UMWA, the coal companies, and the miners themselves. (Elliot)

The health care available in coal-mining communities affects not only working miners, but also their families, other retired miners, and nonmining residents. Left: Miner William Jones at the Cabin Creek Medical Center in West Virginia. (Corn) Top right: A miner's daughter, Phylliss Arthur, at the eye clinic of the Daniel Boone Clinic in Harlan, Ky. (Corn) Retired miner Herman Shupe at the Norton Hospital in Norton, Va. (Murray)

Miners Show Persistent Efforts

Miners have worked hard for their health care. They have fought for better care; they have traveled the long distances necessary to transport the ill and injured to the doctor or hospital; and they have formed coalitions of union, industry, local government, and local business to establish facilities.

Two examples from West Virginia illustrate both the desire of coalfield residents for better health care and the difficulties that they face in obtaining it.

The first is The New River Family Health Center in Scarbro, W.Va., where local residents still recall the days when the only medical care available was from physicians retained by coal companies. UMWA representatives and local residents assessed the health care facilities available in that part of Fayette County and concluded that the small hospital should be supplemented with a health center.

They began work in 1976. A major assist came from the Governor's Office of Economic and Community Development, which helped the residents to plan the project, form the appropriate legal entity, and locate financing. The local group obtained financial assistance from the Robert Wood Johnson Foundation, the Claude Worthington Benedum Foundation, and the Appalachian Regional Commission (ARC). The Fayette County Commissioners (county government) allocated some tax revenues for the project.

The Center opened its doors in the spring of 1979; as of October 1979, the three physicians and staff were treating 80 patients a day. The Center provides physical examinations, screening, immunizations, and other routine care.

The second example is in Cabin Creek, W. Va. In 1973, a miner was injured underground and, after being brought to the surface, lay for 1 hour awaiting transport to a hospital 35 miles away. His colleagues stood around, helpless. They decided at that time to do something — to build a clinic. Countless rummage sales, hot dog stands, and bingo games later, funds sufficient to start construction were in hand. The miners obtained further assistance from the Federal Government, ARC, and Kanawha County Court (county government). Some miners took out personal loans to raise funds, with the organization assuming payments on the loans later.

Two local coal companies, in-

cluding the owner of the mine where the injured man worked, donated heavy equipment and paid the operators to prepare the site. National Guardsmen helped with site construction. Thirty students from the Graduate School of Architecture at Yale University volunteered not only to design the clinic, but to build it, at no charge except for room and board.

The Cabin Creek Medical Center opened in April 1976, and soon found that its fortunes were tied to the ups and downs of the coal economy. The patient load—1,100 for August 1979—drops when the companies lay off miners, and rises when employment goes up. Like so many medical facilities around the Nation, the Center faces a constant battle to match expenses with income, and relies on continued Federal support.

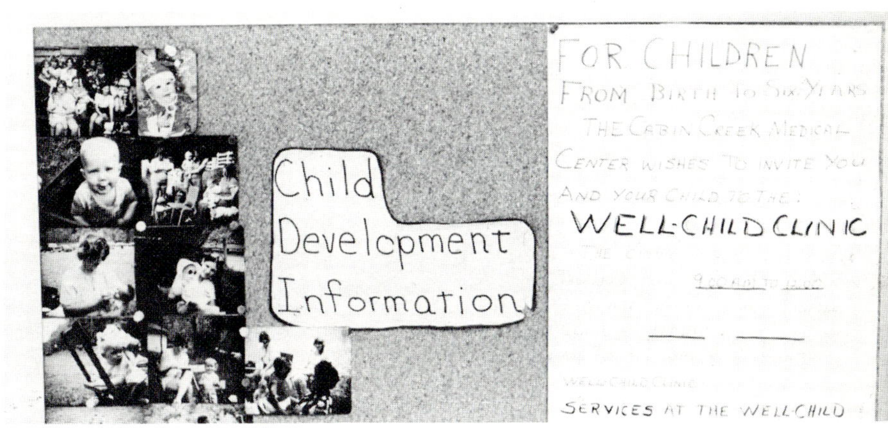

Left: Opening a health clinic in Scarbro, W. Va., took 3 years of persistent efforts. The New River Health Center now serves about 80 patients a day. (Corn) Top right: Following a mining accident in Cabin Creek, W. Va., miners became determined to build their own medical facility. They rallied local residents, located financing, and enlisted college students and others to help. (Corn) Bottom right: Before the Cabin Creek Medical Center opened its doors, residents traveled many miles, over sometimes treacherous roads, to see a physician. (Corn)

Miners' Opinions on Health Care

Not every coal mining community in the United States has a medical center or health clinic. Some lack essential services, and miners and their families living there must travel many miles to obtain treatment.

But many miners do have access to health care on a level that appears substantially to exceed that available in 1946, or ever before.

Miners responding to the Living Conditions Survey express general satisfaction with their health care facilities. Overall, 66 percent say that they are "satisfied" (55 percent) or "very satisfied" (11 percent) with their facilities. Satisfaction levels are higher in northern and central Appalachia and in the Midwest, and lower in the West.

Miners also perceive that their health care facilities compare favorably with those of most Americans. Respondents to the Survey report that 60 percent nationally see their facilities as the same as those of most Americans, and 11 percent see them as better. Again, as with the satisfaction/dissatisfaction levels, miners in Appalachia and the West tend to see their facilities as less comparable. But in all areas a majority of miners perceive their facilities as comparable with or better than those of most Americans.

Financing and Providing Health Care in the Coalfields

Most miners and their families can now choose the doctor, the clinic, or the hospital from which they obtain their routine health care. They are not bound to the medical services provided within their communities. The automobile and third-party insurance contracts provide them with the mobility to seek care in other communities — and sometimes in other States — if they desire. Miners in isolated rural areas face some limitations as to the availability of health care; however, these limitations are the same as those faced by hundreds of thousands of other Americans who live in sparsely populated rural areas.

The system of health care delivery for today's miners and their families is radically different from the system of 30 years ago, or even 2 years ago. It is a system that has evolved over a 120-year period, taking the miner and his family from a health care system that depended on the mining company, to an independent, self-sufficient system, and finally to a system that is interdependent with their community. Over the years, members of American mining communities have experienced both the worst and the best systems of health care that this country has to offer, beginning with the company or contract doctor.

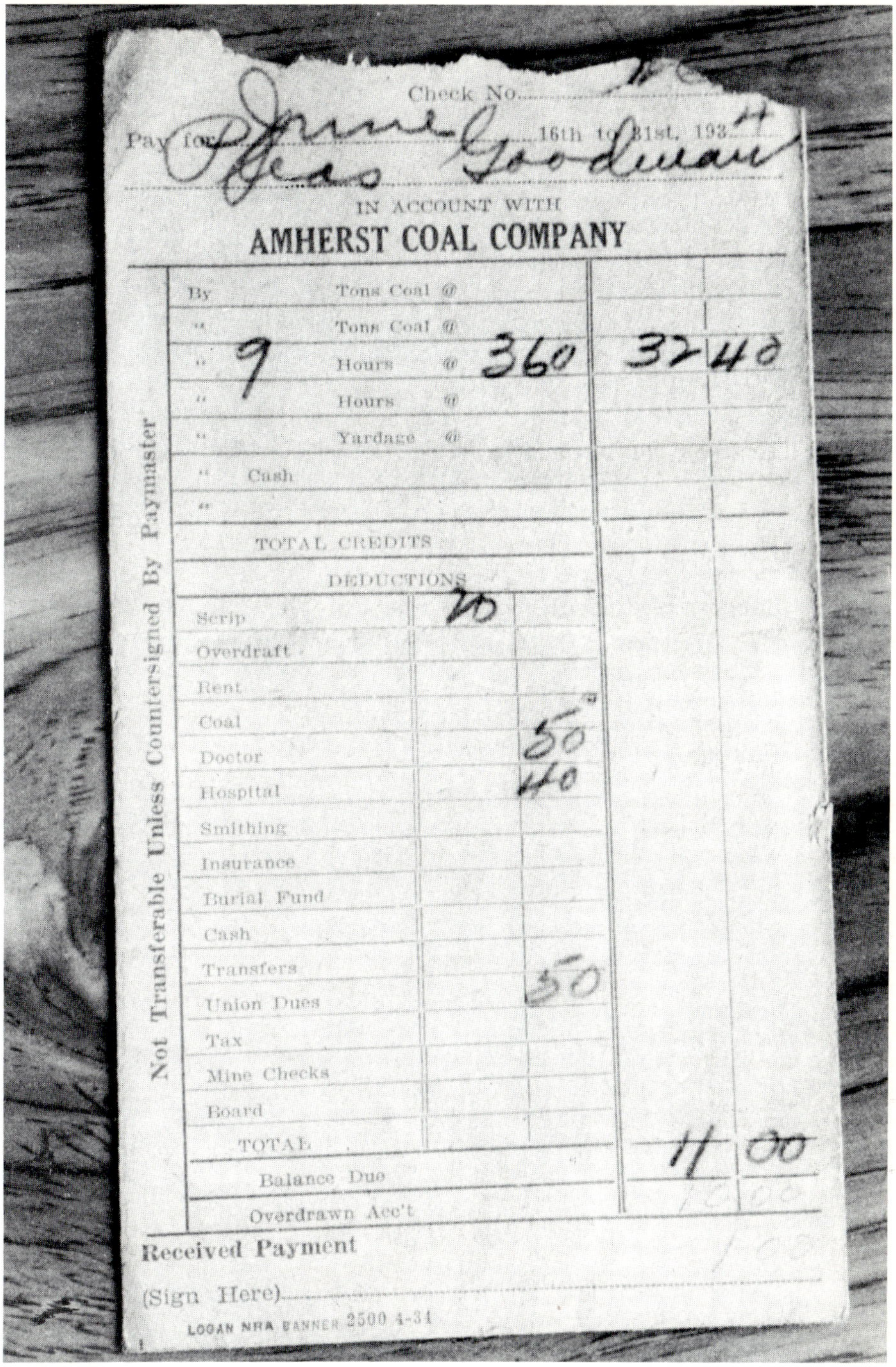

Providing Health Care: The Early Years

In 1842, the George's Creek Coal and Iron Company located in western Maryland, became one of the first American coal mining companies to provide health care services to the miner and his family. These services were available through a prepayment system, a system whereby the miner permitted the company to deduct a fixed amount from his pay each month to support a physician — whether or not he or his family received medical care during the pay period. Under this system, the company hired and compensated the physician, frequently taking a commission as a management fee. The rules and regulations of the George's Creek Coal and Iron Company state that: "Every able-bodied man in the services of the com-

Left: A miner's wife, Mrs. Raymond Weller, at the Cabin Creek Medical Center. (Corn)

Right: Ray Bob Goodman, of Lorado, W. Va., son of a miner, has mined coal for 37 years. He remembers when health care was paid for through company scrip. (Corn)

pany, who will contribute out of his wages a monthly sum of fifty cents for the support of the Physician, will be entitled to medical advice and assistance for himself and family without further charge."

This prepayment system benefited both the physician and the coal company, and generally but not always the miner. Many of the mining communities were located in isolated areas that would not be considered desirable locations for a physician to set up practice. However, the assurance of a fixed monthly income provided the incentive required to bring physicians to these areas.

In coal-mining communities, the company doctor was often the only health care professional available. Frequently he would serve as company doctor for more than one mine in the area, thus enlarging his practice and income. Unfortunately, this meant that the doctor was not always readily available for emergency cases because he was treating patients at another location.

By the 1920s, the system had evolved. The company doctor had become a recognized fixture in the coalfields, similar to company housing and the company store. Fees had risen to an average rate of $1 per month for a single man and $1.50 for miners with families, and a number of companies had begun to arrive at formal agreements with their doctors through written contracts. In addition, the doctors began to charge extra fees for services that were considered to be outside of their normal duties—such as delivery of babies.

The company doctor could provide the miner with routine care, but generally lacked the equipment, and often the knowledge, to treat severe injuries or illnesses. In instances where miners or their families needed surgery, they had to rely on a rather limited number of hospi-

Ca. 1946

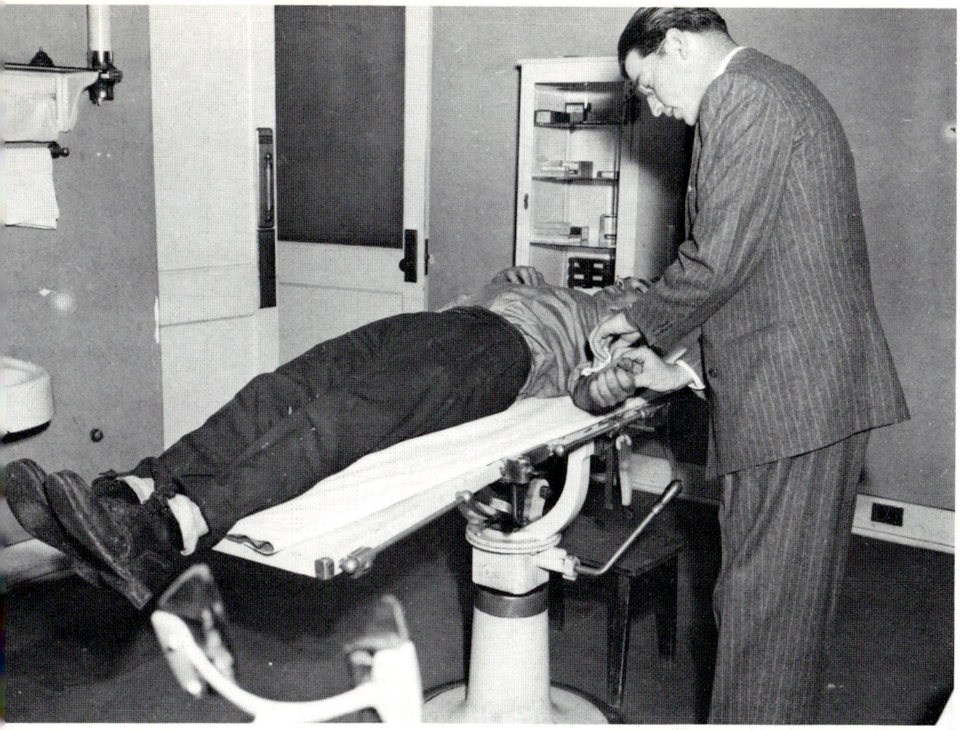

Ca. 1946

Ca. 1946

tal facilities. The 1922 report of the U.S. Coal Commission found that in the 167 towns surveyed, all but 20 had resident physicians, but only 44 had hospitals.

In addition, the Commission reported that:

. . . except for the few camps which had their own hospitals, the journey to the nearest hospital was often very difficult. In remote districts the roads were, for a large part, impassable. Train service was very limited. Unless special trains were made up, injured men might have to wait 12 hours or more and then have a train journey of six or eight hours.

As with the company doctor, most miners were able to prepay for hospital services. Again, the fee was taken out of the miner's pay and disbursed by the company to the participating hospital.

Not all miners and their families were involved in these types of prepayment systems. In mining areas located near urban centers that had a variety of medical professionals and facilities, the miners were responsible for attending to and financing their own medical care. These miners had a degree of choice in the medical care they received; miners in other communities began to demand that choice during the mid-1940s.

For many years, the company doctor was often the only health care professional available to miners and their families. Many of these doctors treated men at the mine site and also maintained offices within the community. Top left: Company doctors in 1946 often served as the local pharmacists. (Lee) Top right: An injured miner receives emergency medical treatment. (Lee) Bottom: Dr. Pierce Martin, a company doctor, in his treatment room. (Lee)

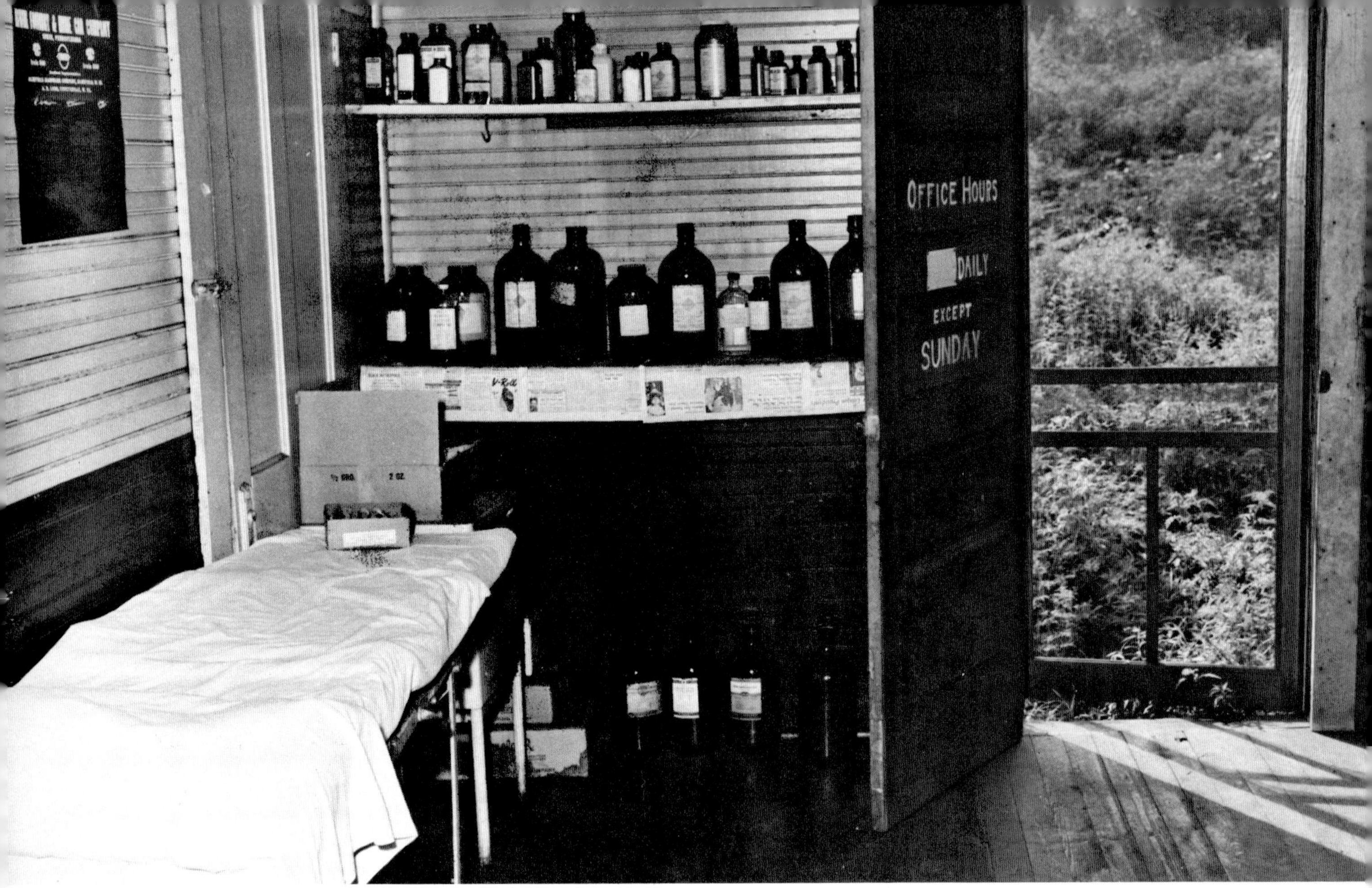

Ca. 194

Reporting on Health Care
The issue of the quality of health care in the coalfields came to prominence during the UMWA contract negotiations of 1946. John L. Lewis, UMWA president, included a welfare package in the contract demands submitted. The union and the mine owners were unable to come to terms and the miners went out on strike when their old contract expired on March 31st of that year.

President Truman immediately ordered the Government to seize and continue operation of the mines in the national interest. On an interim basis, the Government established and financed a health care system for miners. This system became the precursor of the dramatically new health care plan that the UMWA and the coal companies adopted at the end of the strike.

When the new contract was signed on May 29, 1947, miners had won two victories with regard to health issues: first, the coal owners agreed to fund a welfare and retirement fund; and second, the Federal Government agreed to undertake a study of conditions in coal mining areas. This latter agreement resulted from the findings of a study conducted under the direction of Rear Admiral Joel T. Boone, a member of the U.S. Navy Medical Corps.

The Boone Report described a health care delivery system that had not changed much from the time of the 1922 U.S. Coal Commission report. For instance, 59.6 percent of the mines surveyed provided some form of prepayment system for physician health care—a figure that accounted for approximately 70 percent of all working miners. The costs for these services ranged from 75¢ to $3 a month for a single miner, and from $1.20 to $3 for a miner with a family—the average cost being $1.36 and $2, respectively. In 1947, miners earned an average of $63.75 per week.

Aside from the additional expense to the miner, the services provided had not changed substantially. The company doctor was still the usual provider of services and the problems inherent in the system still existed. Many of these doctors continued to serve more than one mine, and the treatment facilities varied. Some offices were well-kept and well-equipped and others were unkempt and ill-equipped.

Thirty years ago, the Boone Report spoke of one of the worst extremes:

. . . 1 physician holds contracts for the medical care of all employees and their dependents at 6 mines, 2 of which were sampled. This doctor receives an average $2 monthly check-off from the mine employees, totaling more than 680 pay envelopes. Addi-

Ca. 1946

tional income is derived from miscellaneous extra charges and the care of private patients. Two of his offices at mine camps were inspected. Both are housed in small, poorly maintained, company-owned buildings.

One of these office buildings, for which this doctor pays the company $12 a month rent, was on a hillside, with a path leading up a moderately steep embankment to the office door. To reach the office, patients must cross railroad tracks and climb the embankment. No steps were provided. The structure was unpainted, and a leaky roof admitted rain water into the waiting room. Equipment in the inner office consisted of an unpadded wooden examining table, an old desk, two unsteady chairs, and a small table. A hydrant near the railroad tracks provided the water supply. Dust blanketed the shelves, where a few bottles of medicine were displayed. Used swabs and soiled bandages were piled high in an open coal crate. The only instruments and diagnostic equipment observed were those that came out of the doctor's bag. Where this office constitutes the medical facility, the physician has been reported to receive an annual gross income of approximately $3,600 from the medical check-off at this one mine alone.

As with the health care provided by physicians, the Boone survey found variations in the availability and quality of health care provided through hospital facilities. Two-thirds of the mines visited provided prepayment plans, a figure that included approximately 75 percent of the miners in the sample.

Again, the monthly cost to the miner varied from 50¢ to $4.80 for a single miner, and from 75¢ to $4.80 for a miner with dependents. On the average, single miners paid $1.67 a month and miners with dependents paid $1.88 per month.

The facilities and services provided through these plans varied greatly. The size of hospitals was found to vary according to the area of the country, with most of the larger hospitals being located in urban centers such as Pittsburgh, Pa. Most miners were not located within geographical proximity of these hospitals and still had to rely on the small local hospitals. Conditions in these hospitals were described as overcrowded: "... single rooms were converted into double rooms ... waiting rooms and porches had been converted into small four- and six-bed wards ... beds were placed in the hallways. In several hospitals ... wards were observed where beds were so crowded together that the space separating them was barely wide enough for a person to walk through."

In the final analysis, the Boone survey estimated that fewer than 20 percent of the mines were served by hospitals with the characteristics of those located in urban centers and that "three-fourths of the hospitals [were] inadequate with regard to one or more of the following: Surgical rooms, delivery rooms, labor rooms and nurseries, clinical laboratories, and X-ray facilities."

The Boone Report focused on many of the problems that the miners had been intimately aware of for many years. With the beginning of a new health care system, the UMWA sought to bring changes that would eventually provide the miners with new choices and medical improvements that had been long overdue.

The Boone Report noted in 1946 that the offices of company doctors varied widely in the sophistication of equipment and supplies and in cleanliness. Left: Although tidy, this office lacks the appearance of a medical environment—grass grows under the front door, and newspapers line shelves. (Lee) Right: This office provides the appearance of a clean, sanitary environment. (Lee)

Pioneering a New Health Care Delivery System

When the new contract was signed in May 1946, the UMWA gained the ability to provide a new system for financing health care for their members and members' dependents. The contract specified that a UMWA-controlled welfare and retirement fund would be established, with financing to come from the coal company owners. Owners were required to support the fund with donations of 5¢ for every ton of coal produced for use or for sale.

The medical program of the UMWA Welfare and Retirement Fund (the fund) began operation in 1949. At that time, it was estimated that a minimum of 50,000 disabled miners, wives, and children were in need of immediate medical care. Word that medical assistance could be received through the fund began to spread through coal-mining communities, and letters such as the following began to arrive at the fund's offices:

I have been a miner for 34 years and on November 5, 1945, I got my back broke and spine injured. I was taken to Logan General Hospital, W. Va., to check my injury and they had already used most of my hospital funds so they sent me back home because they couldn't raise another fund for my care. I went to Huntington, W. Va., to a hospital . . . they couldn't do anything for me there so I came back home. The Compensation Commission wrote me a letter that they couldn't put out no more money for medical and hospital use for me. I have ten in my family and I don't have the money for this operation and I would like you advise what to do, and if I can get help through the Welfare and Retirement Fund to send me to a hospital that has a good nerve specialist. I am in constant pain and burning in my back, legs, and feet and although I'm in good shape for an operation now but if I don't get something done very soon I won't be because I can't stand this pain much longer.

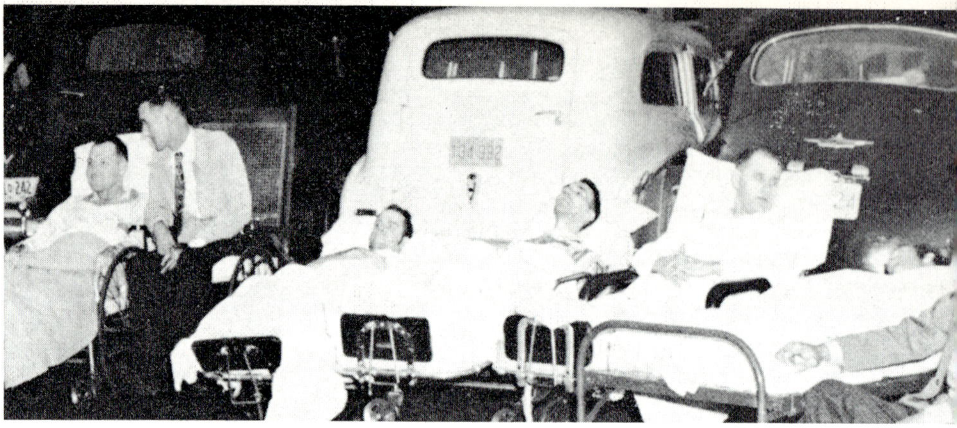

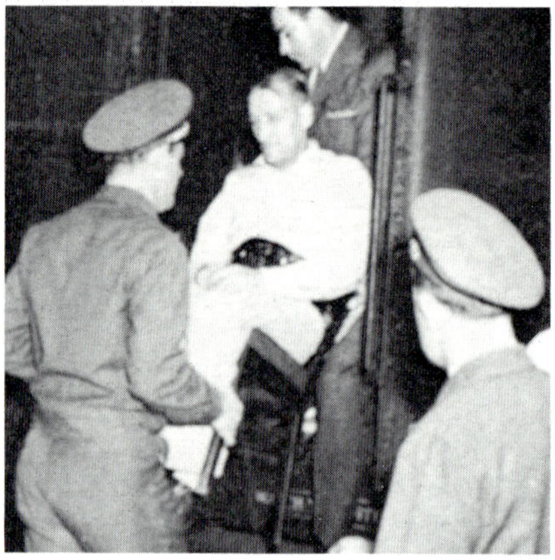

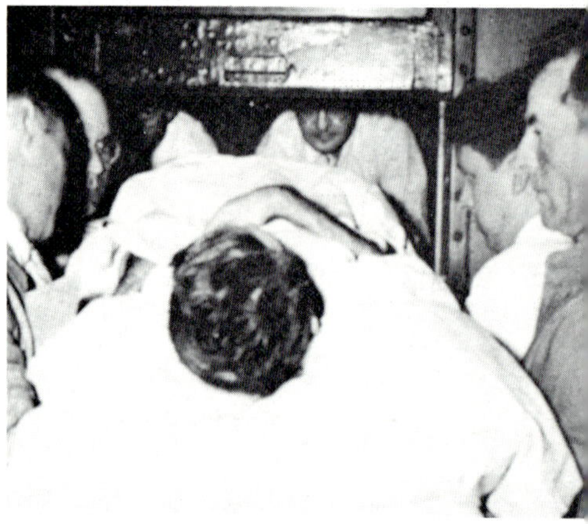

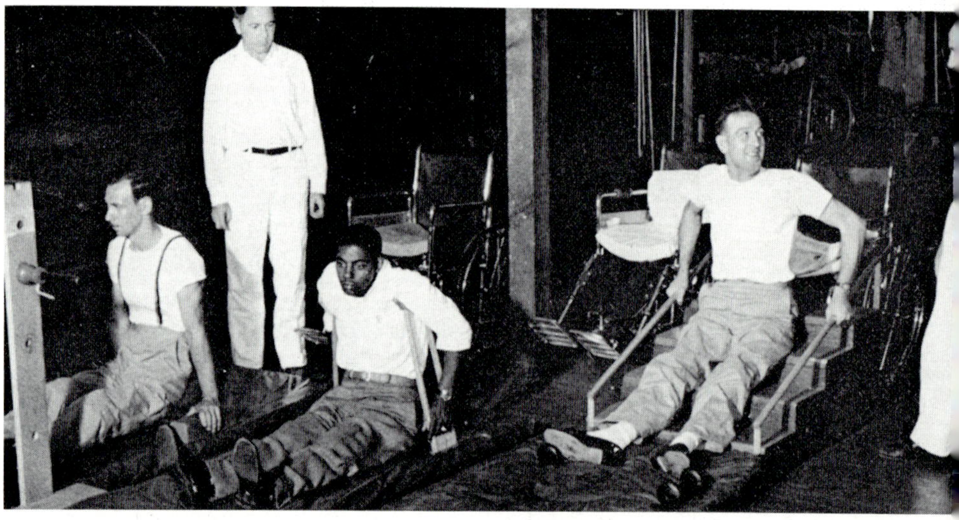

Ninety-seven thousand disabled miners had received medical and rehabilitative help through this program by the end of 1955. Of those, almost 23,000 were able to reenter the work force — some returned to mining and other forms of employment. The aid provided to these miners served as a dramatic beginning to the fund's activities, but it was only a beginning. Many of the miners who returned to their communities still found the same problems with the health care professionals and facilities available to them.

As its first task, the fund set about to locate all these miners and have them attended to by

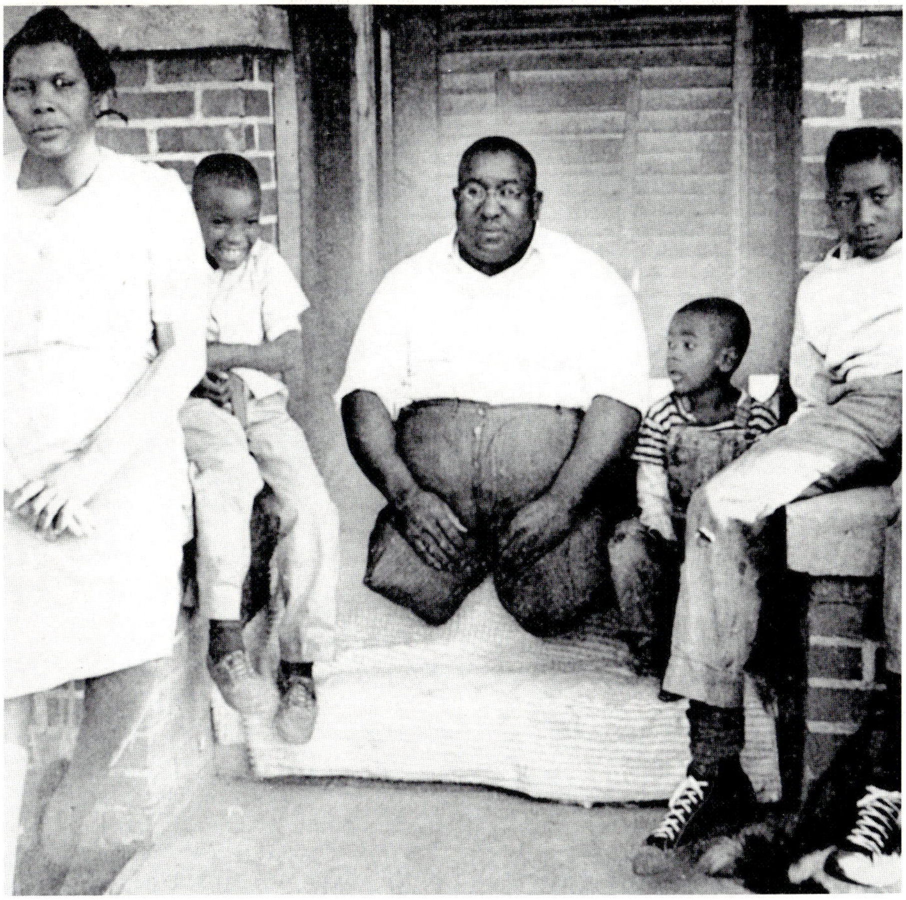

tremities were almost up to his chin. You might ask, is it worth fooling with a person like that? He thought it was. He wanted to live. And we felt we had an obligation. It took 26 surgical procedures and 13 months before we could even start to train this individual... We trained him to walk... on crutches in 90 days, and in control of automatic bladder and automatic bowel. And during the last three months of his stay in the Institute, he ran for sheriff in his county... and he has been the sheriff there for more than three years.

Dr. Warren F. Draper, the fund's first director, stated that the purpose of the fund was "to make good medical and hospital

specialists around the country. Thousands of miners were located, some who had not received medical care for 20 years and who were bedridden under the care of family or neighbors. Dr. Howard A. Rusk, a world-famous expert in rehabilitation, of the Institute of Physical Medicine and Rehabilitation in New York City, described the situation:

These were the toughest cases that anybody ever saw, bar none. You always like to tell about your worst case, but there were many as bad as this one: this man was 40 years old and his back was broken 20 years before. How he survived that length of time, I don't know. When he was found... he... had not seen a doctor in 3½ years. There was not even a wagon road to his house and he was carried down in a sling between two bed poles by friends. The man had 11 bed sores from the size of a plate to the size of a dollar; stones in both kidneys and his bladder, and his lower ex-

The country watched as the new UMWA Health and Welfare Fund brought thousands of debilitated miners out of the coalfields for rehabilitation. Many of them had been bedridden for more than 20 years. By 1955, 97,000 disabled miners had been transported to hospitals and rehabilitation centers by special trains and ambulances. Top and center left: Disabled miners wait to be placed aboard special trains bound for rehabilitation centers. Bottom left: Miners undergoing therapy required to strengthen arm muscles as a prelude to learning to walk with the aid of crutches. (UMWA Health and Retirement Fund)

Sam Harris was one of the first miners to be helped by the new UMWA Health and Welfare Fund. In 1938, his legs were crushed in a mining accident—both legs were subsequently amputated. As a beneficiary of the Fund program, Mr. Harris was sent to the Kessler Institute of Rehabilitation, where he underwent physical conditioning and was fitted for artificial legs. Top right: Mr. Harris and his family before receiving assistance from the fund. Bottom right: Mr. Harris, 95 pounds lighter and able to walk and work again with the aid of artificial legs and crutches. (UMWA Health and Retirement Fund)

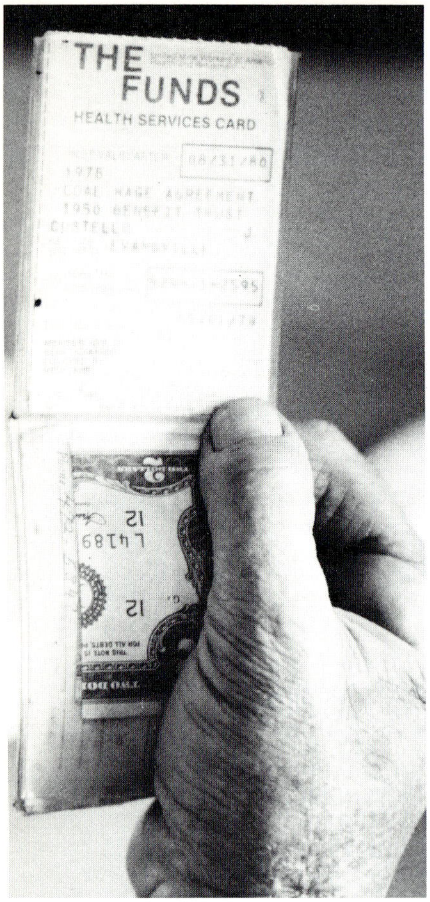

care possible at reasonable cost." Medical services authorized by the fund included:

- Hospital care, including special duty nursing, nursing home, and rehabilitation services when necessary;
- Physician services, including hospital services, group practice, mental illness diagnosis, and short-term therapy;
- Drugs, including those used during hospital confinements and nonemergency situations, but not expensive drugs needed for long-term medical problems; and
- Appliances and prostheses not available through other sources. Ordinary dental care was not funded.

Union miners were provided with a card that they presented to their physician to receive medical care. The physician was then responsible for submitting an invoice to the fund. If hospitalization was required, the physican simply provided the beneficiary with a note that was turned in to the local union for a hospitalization slip. This slip was then presented to the hospital and the hospital invoiced the fund.

For the first few years that the fund's medical benefits program operated, miners and their dependents were able to obtain medical services from the physicians and hospitals of their choice. This system sometimes led to abuses by physicians — unnecessary surgery was frequently cited as an example — and the fund was forced to establish a list of physicians who would be reimbursed for providing services. In addition, miners were required to obtain second opinions when surgery was suggested, except in emergency situations.

Miners took pride in having UMWA health cards. The cards offered a kind of security that had never existed in the coalfields before — the knowledge that one could always get and pay for medical assistance. In addition, the miners began to see improvements that their health benefits system brought to their communities.

During the 1950s and 1960s, many new medical facilities and medical professionals had become established in coal mining communities. Often this increase in the availability of medical care was a direct result of the fund. However, several things soon became evident: first, finances provided by the fund were not only supporting services for miners and their dependents, but also they were often indirectly paying for health care for the entire community. For example, contemporary observers reported instances of miners loaning their health cards to relatives or friends who needed medical treatment but could not afford to pay for it.

Second, the system became heavily overused. Miners and their dependents frequently sought health care unnecessarily. There were allegations that doctors abused the system by exaggerating the condition of patients and continuing them unnecessarily long in treatment. There were also allegations that doctors continued to charge the system for office visits after discharging the patient.

Finally, a series of wildcat strikes and other factors seriously diminished the royalties anticipated by the fund. By the mid-1970s, the system had become financially overburdened, and after almost 30 years began to crumble under its own weight.

Copaying for Medical and Hospital Benefits

During the fall of 1977, miners began to pay for a percentage of their medical care costs once again. The fund had become insolvent. (In 1974, the fund had been separated into health funds and pension funds.) Although the fund continued to receive payments from the coal companies, these payments did not cover the costs of providing medical and pension benefits to its beneficiaries.

The new contract signed by the UMWA and the Bituminous Coal Operators' Association in the spring of 1978 included a new health benefits system. Although the 1978 contract stipulates the benefits to be provided to working miners and their families, each mine operator is responsible for purchasing health insurance for its employees and their families.

The system now uses standard medical insurance carriers, such as Aetna, Blue Cross, Prudential, etc. It is basically the same as that adopted by all major industries, covering the majority of workers in the United States.

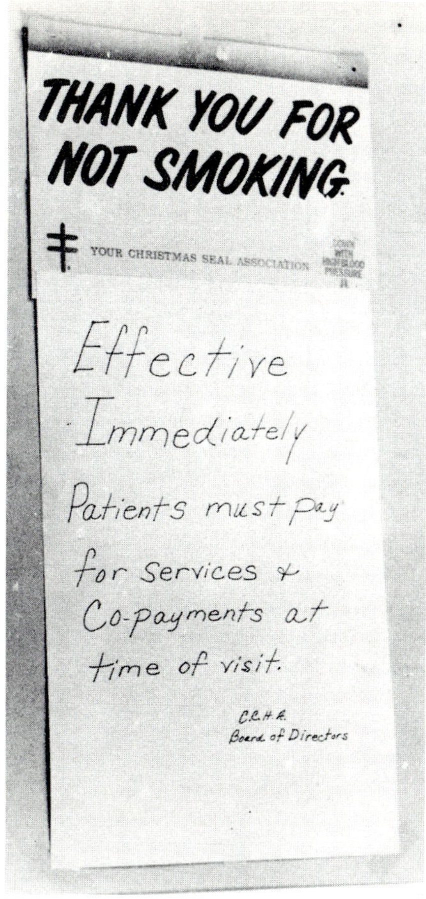

As part of the 1978 contract, miners who had been copaying for medical expenses during the previous 6 months were permitted to submit their invoices for refunds. The coal operators continue to pay royalties into the UMWA Health and Retirement Funds, largely to pay for the fund's outstanding obligations to older miners for health and pension benefits and for pensions for working miners.

Today's miner and his or her dependents pay \$7.50 — up to a maximum amount of \$150 per family per year — each time they make an office visit to their physician; no such copayment is required for hospitalization. Some may feel that they have taken a step backward, but the frequency of visits to the doctor for unnecessary reasons has decreased. As one miner's wife explained: "The union cards used to pay for anything. I'd take my Daddy to the doctor when he was sick and it would be running over and you'd have to wait 'til you could get in to see the doctor. People would come in for any little old thing because they knew it was paid for. Then when they had the last contract, they stopped that. Now you have to pay so much each visit. Now it doesn't take hardly any time when we go in."

Both union and nonunion miners have gained better health care from the systems developed for the UMWA. One benefit is the increased number of medical care providers available in the coalfields. Another — for nonunion miners — involves the view of operators of nonunion mines that they must offer a health care plan similar to that of the UMWA in order to attract workers.

Far left: Although working miners no longer use this type of identification card to obtain medical care, pensioners and disabled former UMWA miners, such as John Costello, from Helper, Utah, do. (Corn) Center left: UMWA health cards not only served to finance visits to the doctor and to the hospital, but also paid for medication. (Murray) Right: At the Cabin Creek Medical Center in West Virginia, and many other clinics in coal-mining communities, miners and their families now pay the partial cost of each office visit before receiving medical care. (Corn)

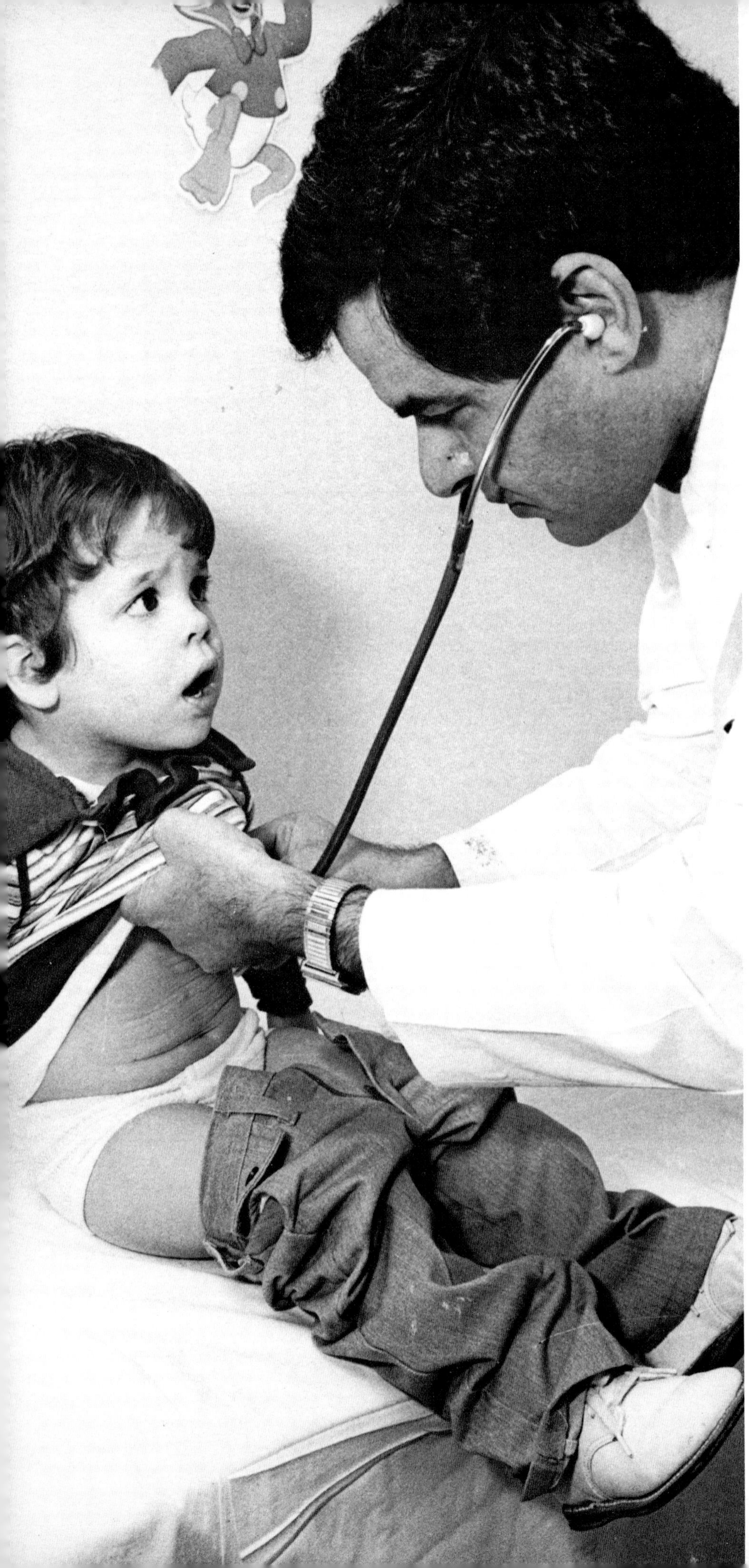

Health Care Providers

When the Administrator of the Campbell County Memorial Hospital in the coal boomtown of Gillette, Wyo., decides that his hospital staff needs an additional physician, a community effort begins. Recruitment of candidates is assisted by Western Health Systems, Inc., a federally subsidized program serving Montana, South Dakota, and Wyoming, but the final selection process involves the hospital Administrator, the hospital staff, and community residents.

Before physicians are formally offered staff positions, they are invited to visit Gillette with their families. While his or her spouse tours the town — including schools, shopping facilities, and available housing — the physician tours the medical facility and meets with the hospital Administrator and members of the staff. Interspersed with these meetings are receptions and dinners with the Mayor and other civic leaders. The process takes about 2 days, and the hospital pays for the trip.

Once the hospital selects a physician, it rents office space to the doctor and guarantees a first-year gross income of $80,000. Usually, when the physician enters his office on his first day of practice, even as the final touches of furnishing are being completed, his waiting room is full of new patients.

The recruitment effort is so successful that, despite chronic physician shortages in many places in the United States, Campbell County Memorial was not looking for any doctors in 1979.

The quantity and quality of physicians and medical facilities are an important concern of miners and their families. The physicians and health care facilities found in coal-mining communities today are a far cry from those in 1946. At the heart of the changes are the increased

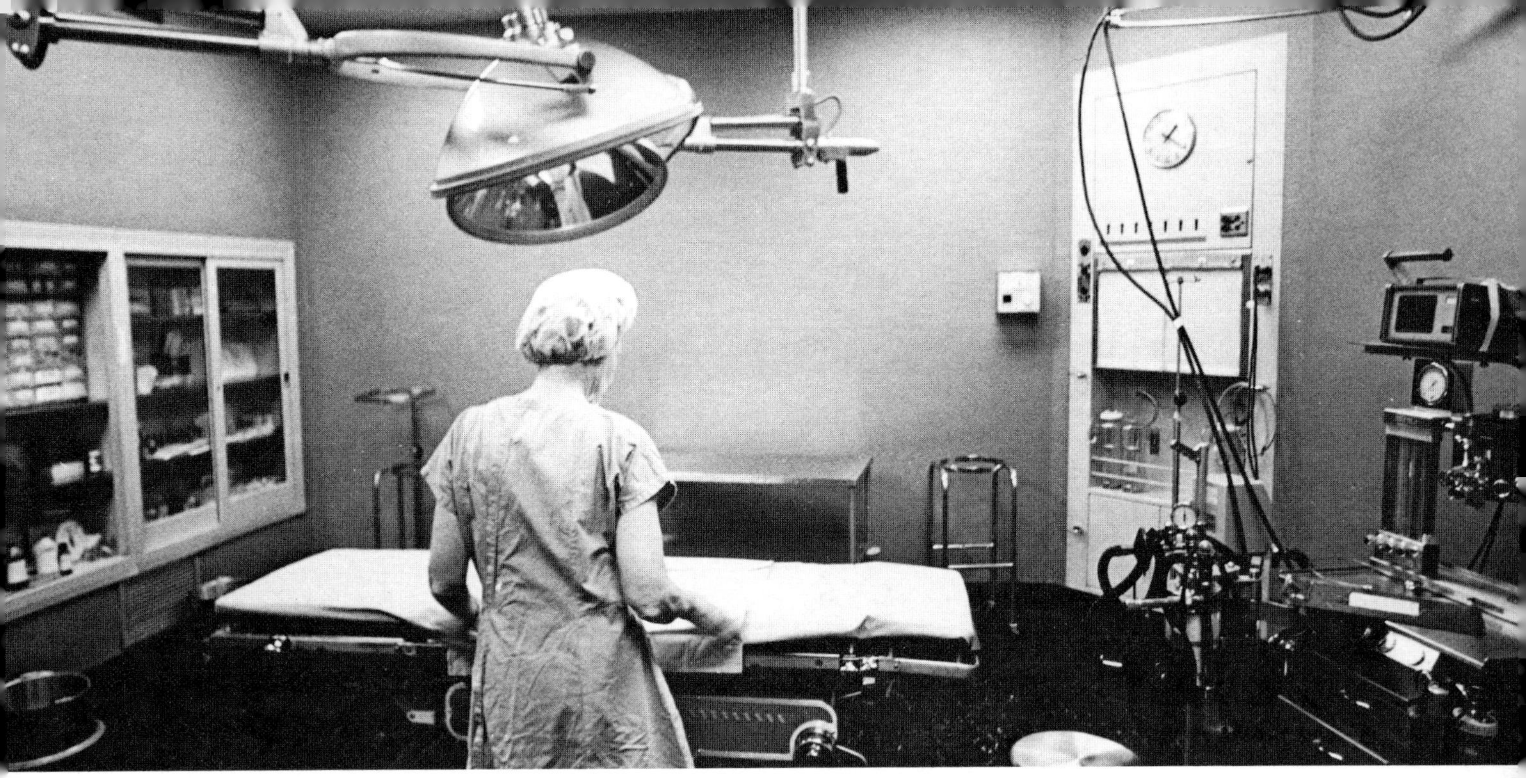

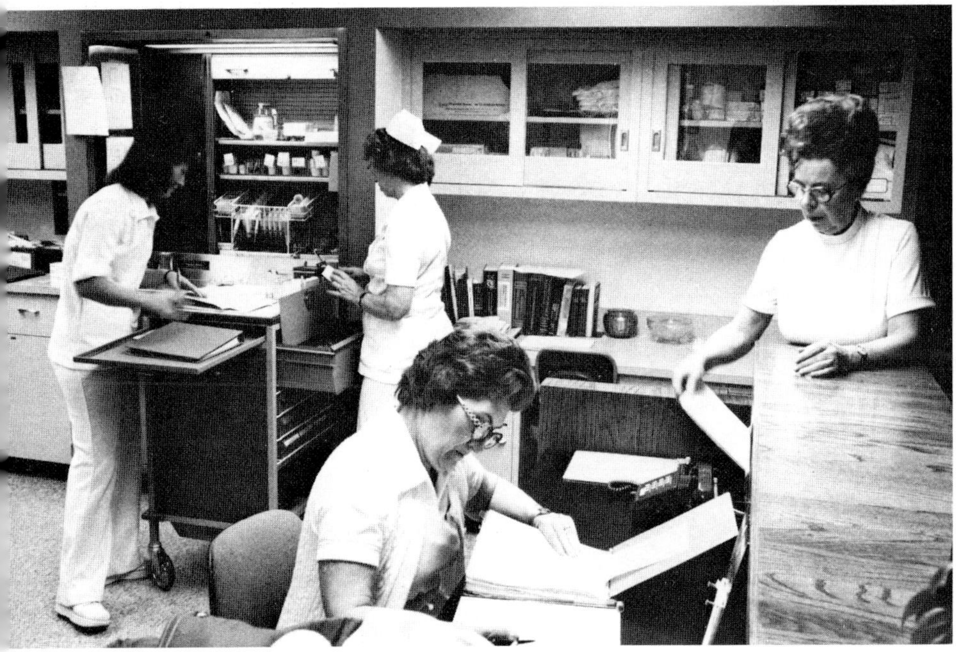

number of physicians, hospitals, and clinics, and the increased role that miners have begun to play in selecting where and by whom they and their families are treated.

Examining and Assessing Health Care Providers

Assessing the quality of medical services available in any community involves a subjective element. Although it is possible to count physicians, facilities, and up-to-date equipment, the practice of medicine continues to be an art — not an exact science. The successful practice of medicine relies heavily on the physician's skill in communicating with the patient, and on the proper assessment of the patient's condition. Two doctors may examine the same patient and decide on different courses of treatment — each of which may be appropriate.

This study does not attempt to

The quality of health care depends not only on the quality and availability of physicians, but also on the quality and availability of nurses, health care facilities, and equipment. Left: A miner's son, Jonathan Myers, being treated by a pediatrician, Dr. D. J. Malie, at the Jackson Clinic in Jasper, Ala. (Simmons) Top right: An operating room at Harlan Appalachian Regional Hospital. (Corn) Bottom right: A nurses' station at Harlan Appalachian Regional Hospital in Harlan, Ky. (Corn)

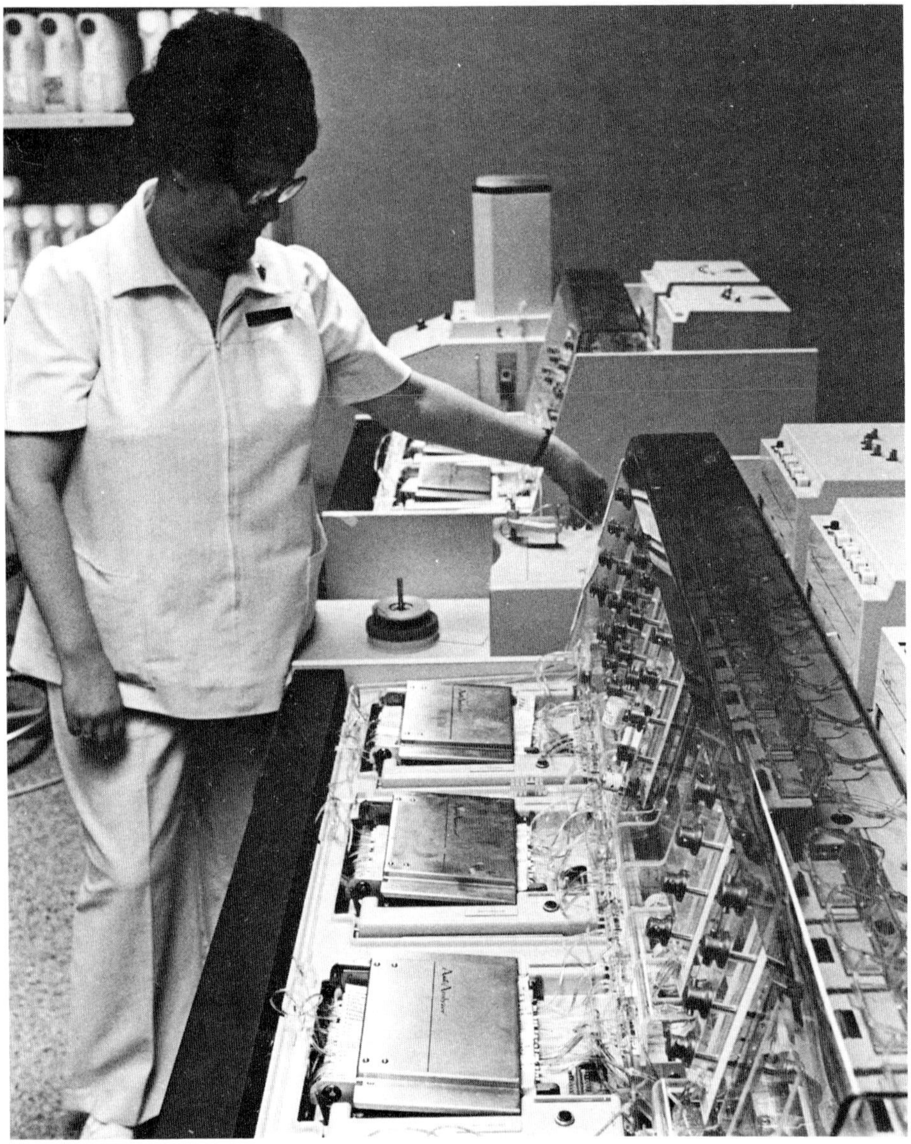

assess the quality of health care from that patient-level point of view. Rather, it concentrates on the major factors — the presence and numbers of health care providers and facilities.

To present this picture, the study analyzes 169 counties that produce the most coal and have the highest percentage of coal miners in current employment. The counties are located in 17 coal-producing States.

The statistics used to examine medical care in these counties were obtained from three sources: American Medical Association (AMA) publications *Distribution of Physicians in the United States 1970* (the first year county-by-county data were provided) and *Physician Distribution and Medical Licensure in the U.S., 1977;* U.S. Department of Energy (DOE), *Regional Profile: Energy Impacted Communities Region VIII;* and various statistical studies of the U.S. Department of Health, Education, and Welfare (HEW). The AMA's 1977 analysis and the DOE study were published in 1979. They include the most recent statistics available on this subject.

The trends identified from these materials indicate overall positive improvements in the availability of health care providers in the coal-mining counties of the United States. There remain areas, however, where substandard care is prevalent.

Overview of Physician Availability

In 1970, 8,743 non-Federal physicians provided primary care to residents in the 169 coal-producing counties surveyed. By 1977, that number had risen to 11,058 — a 26.5 percent increase.

The largest increases are in specialized medicine. The general practitioner (GP), at one time the only physician found in coal counties, is less prevalent. In 1970, residents in 20 percent of the 169 coal counties relied on GPs alone for their health care. In 1977, that figure had fallen to only 8.3 percent.

Many coal-mining communities today have internists, family practitioners, thoracic surgeons, orthopedic surgeons, and the like, to provide for their medical needs. Between 1970 and 1977, the number of GPs in coal communities decreased by 12.9 percent. During that same period, the number of doctors with medical specialties increased by 44.9 percent, the number of surgeons increased by 23.6 percent, the number of other specialists — such as occupational medicine — increased by 56.6 percent, and the number of physicians in hospital-based practices increased by 49.9 percent.

Increases in the number of physicians were not uniform. They vary with the population of the individual coal county, with the largest increases occuring in counties that include or are contiguous with metropolitan areas.

Physician/Patient Ratios

The current national average physician/population ratio is 1.3 physicians per 1,000 in population. The ratio of physicians per 1,000 in population is often used to estimate health manpower needs. In 1970, the physician/population ratio in mining communities with less than 50,000 residents fell well below the national average. During the same year, counties with larger popu-

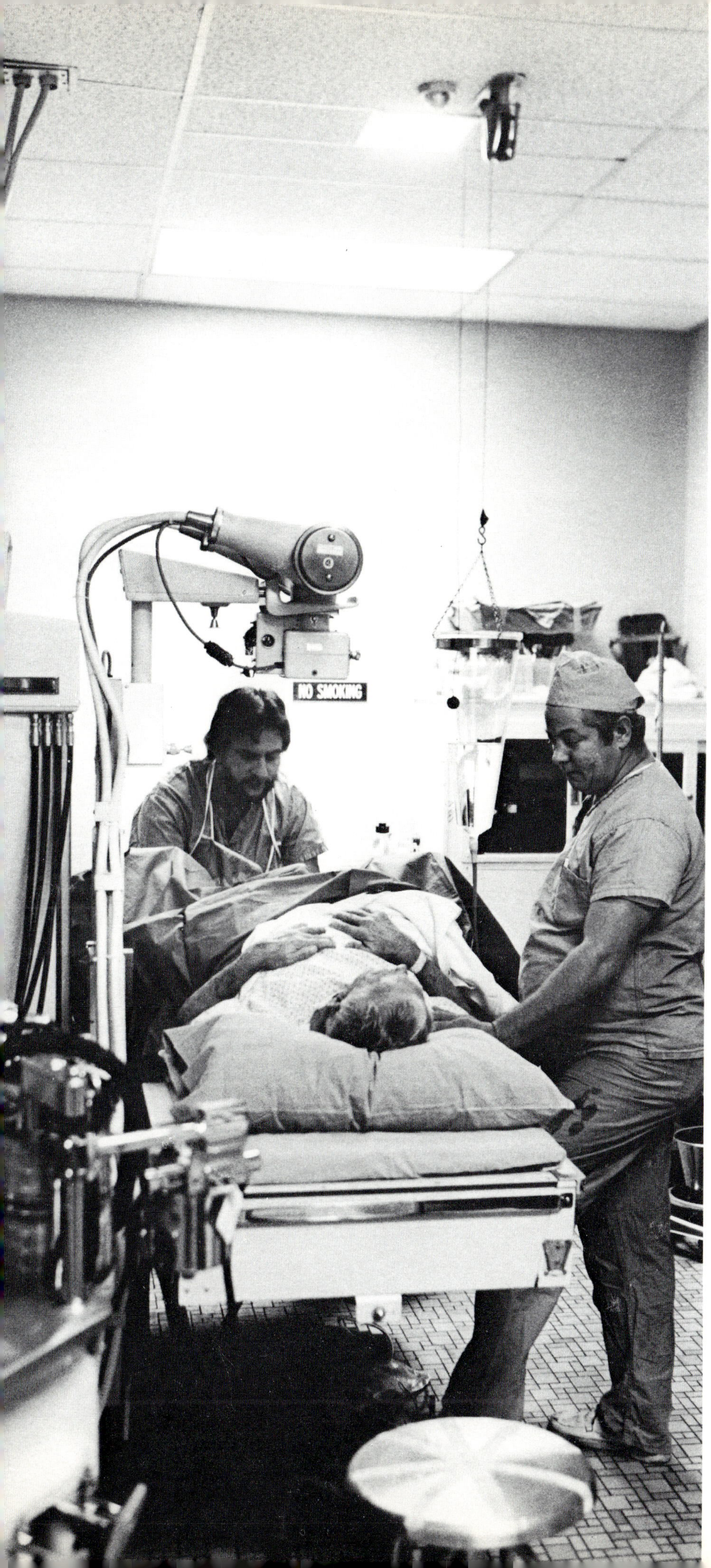

lations exceeded the average. In 1977, many of the same communities continued to show shortages, but improvements were noted. For instance, 50 of the 169 counties studied have populations of from 25,000 to 49,999. The physician/population ratio in these counties was 0.57 in 1970, but it rose to 0.68 in 1977.

Hospitals

As with physicians, hospital facilities in coal communities have changed over the years. Between 1970 and 1977, a net gain of one hospital brought the total number of hospitals for the 169 counties to 320. This figure is, however, deceptive.

During that 7-year period, a number of hospitals closed but many new facilities also opened. For instance, in counties with populations of up to 10,000, six hospitals were closed, leaving the counties represented with only 11 hospitals in 1977. Although a loss of six hospitals is shown, the total number of hospital beds increased by 69, and the average number of beds per hospital rose from 29.5 to 51.8. This would indicate that a number of the smaller facilities were closed and replaced by a lesser number of larger hospitals.

According to a study by Appalachian Regional Hospitals, Inc., the ratio of hospital beds to population in rural areas should approach 3.5 per 1,000 residents if a community is adequately served. In 1977, most of the 169 coal counties surveyed had hospital bed/population ratios that equaled or exceeded that estimate.

Left: The laboratory at Harlan Appalachian Regional Hospital. (Corn) Right: A miner being prepared for surgery at Harlan Appalachian Regional Hospital, Harlan, Ky. (Corn)

Clinics

Located on the edge of the rugged Cumberland Plateau, Petros, Tenn., is a community of about 1,200 persons. It was once a large underground mining center, but as mining became mechanized the number of jobs available declined. Many miners left, and with them went the community's lawyers, shopkeepers, and doctors.

By 1970, the community lacked even the basic requirements for providing adequate health care. The residents of Petros decided to take the responsibility for improving the situation into their own hands. After 2 years of analysis and planning, they formed the Petros Health Council, Inc., and began efforts to raise funds and build a community health clinic. Bake sales, suppers, film showings, gospel sings, and raffles eventually produced enough funds to begin construction of the facility; most of the labor was donated by local people.

The clinic opened its doors in 1973. It is now staffed by a full-time physician, a nurse practitioner, a receptionist, and a clinic aide. The physician and nurse practitioner, both provided by the National Health Service Corps, conduct minor surgery, examinations, prenatal care, laboratory work, general medical care, and home visits. Approximately 500 people receive medical care from the clinic each month.

Although Petros is not one of the more active coal communities, its fight for quality medical facilities and care is the same fight that many other coal-mining communities have undertaken. Mining communities have banded together time and time again to assess individual community needs, obtain financing, construct facilities, and recruit medical professionals. They have often obtained aid from the UMWA, Federal and State gov-

ernments, and sometimes the coal companies themselves.

The clinic is one of the most important health care facilities in coal-mining communities — particularly in the rugged, rural regions of Appalachia. It is these facilities that provide care to many people who would otherwise have to travel hours for the treatment of even the most basic of medical problems.

There are hundreds of clinics in Appalachia. Most are small, often with only one or two physicians and a few nurse practitioners to carry the patient load. These clinics generally provide basic family practice medical care.

There are also large clinics — clinics with upward of 15 physicians on staff, and a full complement of nurses, facilities, and diagnostic equipment. These larger clinics are, however, usually found in the larger, more heavily populated communities, where hospital facilities are also available. They often serve as a medical focal point for the community, providing a full array of physician specialties that would not normally be found in these rural settings.

Top left: The Petros Clinic, in Petros, Tenn. Built in 1973 through the efforts of the community's residents, this facility provides health care to 1,200 people. (Corn) Bottom left: The waiting room of the Petros Clinic. (Corn) Top right: The Twin Cities Medical Clinic, in Huntington, Ark., was opened in 1977 by Dr. A. P. Venturina. Of the approximately 150 patients seen at the clinic each week, 20 percent are miners. (Simmons)

Health Care Providers in Six Coal Counties

Statistics tell only part of the story of health care in coal counties. For a more detailed picture, it is necessary to take a close look at individual areas — where they are located, and what factors influence local health care decisions.

This more detailed picture also provides the opportunity to compare those locales with similar-sized, non-coal-producing counties.

For the purpose of assessing physician distribution, the AMA classifies every U.S. county by its resident population and proximity to a metropolitan area. All of the 169 coal counties surveyed fall into eight of these county groups. Six illustrative counties are described below.

Moffat County, Colorado

Forty-two hundred people lived in Moffat County in 1970. By 1979, as the coal boom continued to progress, that number had swelled to 11,800 — a 280 percent increase. More than 75 percent of the county population resides in Craig, the county seat.

With the exception of one nurse and six emergency medical technicians, all county health providers — both facilities and professionals — are located in Craig. Health care facilities include: one 30-bed hospital, one mental health clinic, an alcohol detoxification center, and two nursing homes. All of Craig's 14 physicians have staff privileges at the hospital, although only a few conduct their practices within the hospital's facilities.

Memorial Hospital is located in the residential area of Craig, behind the school administration building. The dark red brick facilities represent two eras of medicine. The front lobby-reception area is sunny and well lit. The floor is covered with burnt red wall-to-wall carpeting,

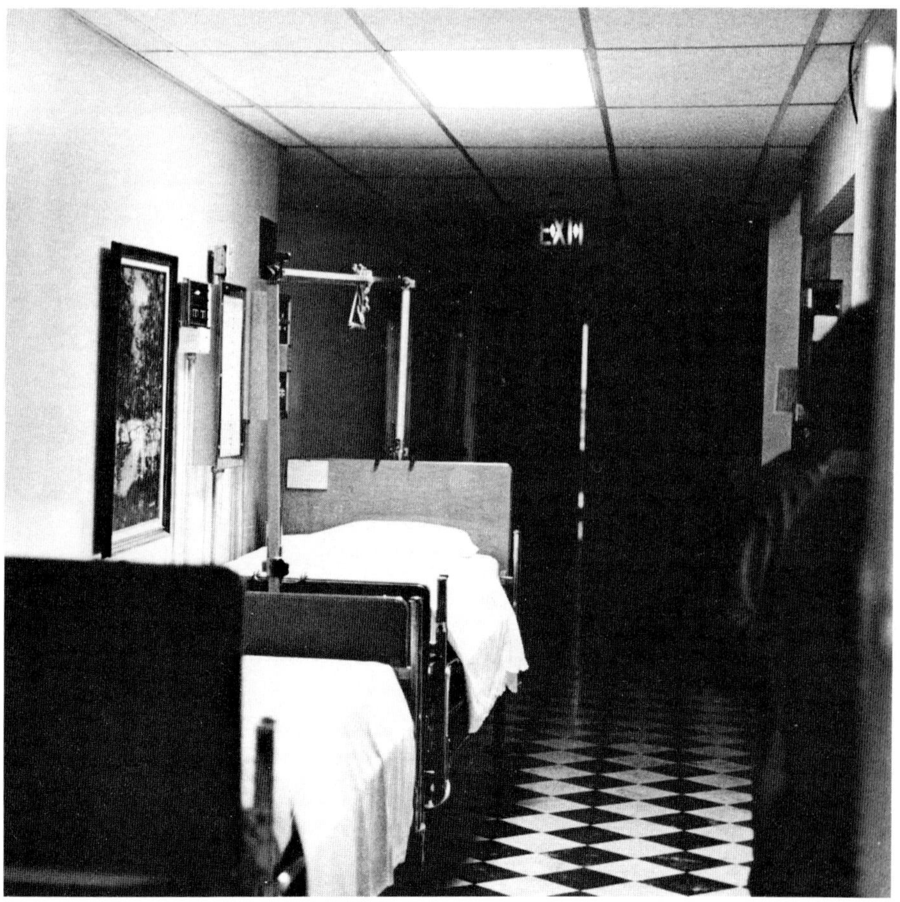

free-standing plants dot the seating areas, and a geometric weaving hangs on the wall. Yellows, pinks, reds, and oranges are used to brighten examination rooms, cabinets, chairs, and file drawers. "No smoking" signs appear at every corner. This is the new half of the hospital. A walk through a connecting door opens up a different world.

The old wing of Memorial Hospital houses patients. It is dark, lit by low-watt bulbs recessed into a white acoustical tiled ceiling. Brown water stains spot the ceiling. In the halls are collapsed wheelchairs, cleaning carts, and a lone elderly patient shuffling along using an aluminum walker for support. At the nurses' station are five nurses crowded together in a limited space.

Construction and renovation of Memorial Hospital were begun in 1977. The new building, which contains the admitting area, X-ray facilities, an emergency room, and administrative and physician offices, has been completed. Work on a second building, which will house 13 new beds, a second operating suite, and new obstetrical, physical, and respiratory therapy areas, is now under way.

There were only four physicians in Craig 5 years ago. Today, the 14 physicians who serve Memorial Hospital provide specialties in internal medicine, general surgery, radiology, and family practice. In addition, the hospital staff is augmented by

Top: The new lobby of Memorial Hospital in Craig, Colo. (Berger) Bottom left: The corridor of the "old" wing of the hospital. (Berger) Bottom right: Outside the pharmacy in the "old" wing of the hospital. (Berger)

specialists who travel to Craig on a regular basis: once a week an orthopedic surgeon comes from Steamboat Springs, and once a month specialists in ophthalmology, ear/nose/and throat, plastic surgery, urology, and neurology come from Grand Junction.

The hospital laboratory is staffed by four full-time technicians. They have the equipment to conduct basic diagnostic work, but all tissue and pathology samples are sent to Denver for analysis.

In 1978, when DOE surveyed health care providers in Craig, the town supported nine physicians, three physician assistants/nurse practitioners, three dentists, two optometrists, and five other health care professionals. These totals help to paint a picture of continuing growth in the availability of medical services in Craig over the past 10 years.

When the AMA assigned counties to groupings in 1977, Moffat fell into the number one group — nonmetropolitan counties with populations of up to 9,999.

At that time Moffat had six physicians: three general practitioners, one surgeon, and two specialists. These data showed a marked change from the 1970 statistics when Moffat had only three physicians, and all were in general practice.

Moffat County compares favorably with other same-size coal counties. In 1977, 17 of the Commission's sample coal counties were assigned to the AMA's group one. Of these counties two — Putnam, Mo., and Oliver, N. Dak., — had no physicians; six had no hospital. However, on the average, 0.61 physicians and 3.6 hospital beds per 1,000 population were available in these coal counties; in Moffat, the figures were 0.8 and 4.2, respectively. Nationally, the average was 0.43 physicians, and 4.1 hospital beds per thousand in counties of this size.

Campbell County, Wyoming

Campbell County stretches 101 miles from north to south and 46 miles from east to west. It has a population of approximately 21,000, many of whom are recent transplants to the area. The county is rich in coal, oil, and uranium, and its population growth is expected to reflect the increasing demand for these resources in coming years.

Gillette is the largest town in the county. It has a population of about 13,000 within the city limits and an additional 4,000 in the surrounding areas — 81 percent of the county's population. The only hospital in the county is located in Gillette.

Campbell County Memorial Hospital is a red brick, white trimmed, single-story building. The 31-bed facility was built in 1951 to serve the county's 4,800 residents. The hospital shows signs of overcrowding. All available space is taken up: storage cabinets have been built into hallways.

In 1976, there were seven physicians practicing at the hospital. Today, there are 23, including 6 internists, 2 obstetricians/gynecologists, 2 pediatricians, 4 family practitioners, 2 surgeons, 1 pathologist, 1 radiologist who is also the hospital Administrator, 1 specialist in occupational medicine, 1 anesthesiologist, 1 orthopedic surgeon, 1 emergency room physician, and 1 ophthalmologist.

Although small, the hospital has the diagnostic and treatment capabilities of a 100- to 200-bed hospital. It maintains a 70 to 80 percent occupancy rate, and its physicians generally earn gross incomes in excess of $80,000 per year.

Within the past 4 years, the hospital has expanded its basic plant, but has not increased the number of beds. That increase

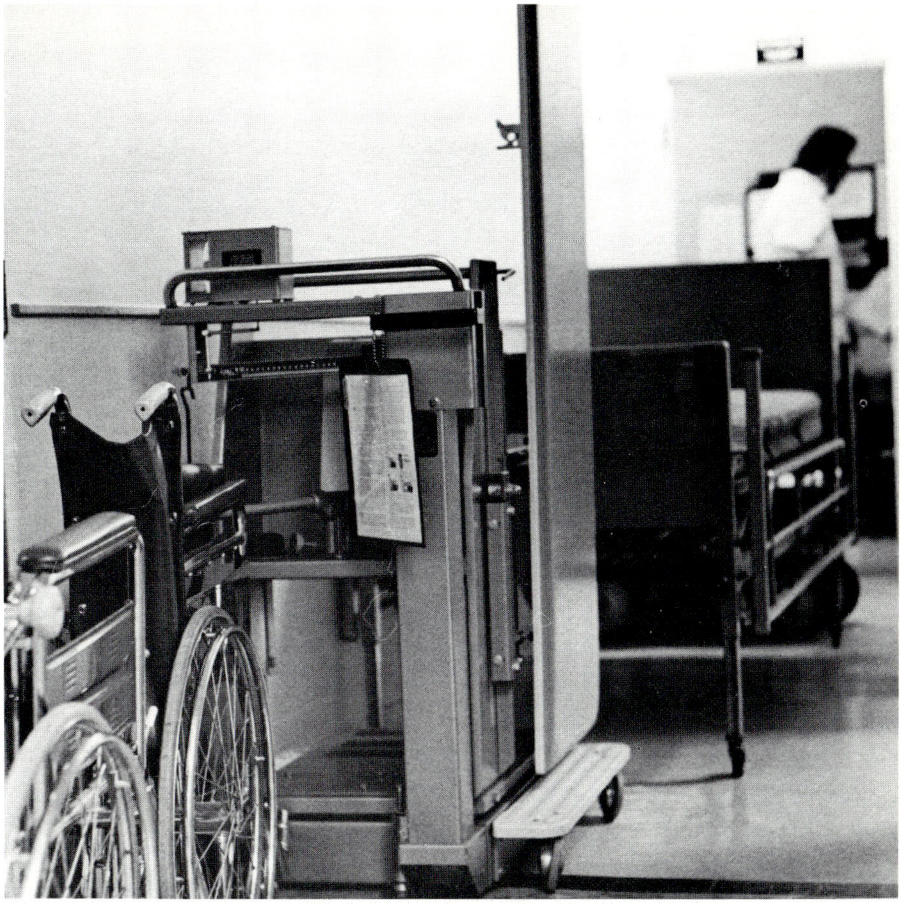

will come in mid-1980, when Gillette's second hospital will be completed. The new facility will provide 55 additional beds.

All of Gillette's medical professionals have their offices either within the hospital or in the adjacent prefabricated medical buildings. The Health Services Research Institute is located in one of those buildings.

Four of Gillette's physicians — two family practitioners and two obstetricians/gynecologists — and two physician aides are based at the clinic. The town also supports 2 optometrists, 5 dentists, 5 public health nurses, 11 emergency medical technicians, and 4 psychologists.

The hospital Administrator says that although he is turning away applications from physicians, the hospital has had difficulties in obtaining an adequate supply of nurses. In addition, he says that until recently, at least half of Gillette's population was traveling to other facilities for medical care because they did not have confidence in Gillette's hospital. Today, he says, that image is changing because of increased funding, the hiring of more and better doctors, and an improving word-of-mouth reputation.

The current physician level in Gillette is almost double what it was for the whole county only 2 years ago. In 1977, the AMA assigned Campbell to its number two county group — nonmetropolitan counties with populations ranging from 10,000 to 24,999 inhabitants. At that time, 12 physicians and one hospital served the county's estimated 13,700 residents. Of the physician population, two had surgical specialties, and two had other specialties. None of the physicians was in a hospital-based practice. The trends of growth and specialization in Campbell County are further shown by comparison with the AMA's 1970 data. At that time the county had only eight physicians; half were GPs.

Comparisons with same-sized coal counties and with the Nation give a fuller picture. Of all AMA counties in this category, 59 meet the criteria as coal-producing counties. In those counties, there are an average of 0.49 physicians and 3.36 hospital beds per 1,000 of population; in Campbell County, the figures are 0.88 and 2.26. Nationally, there are 0.53 and 3.88 hospital beds per 1,000 of population in counties of this size.

All medical professionals and facilities in Campbell County, Wyo., are located in Gillette, at the hospital and the adjacent medical complex. Left: Outside Memorial Hospital. (Berger) Right: The storage of wheelchairs and other equipment in the corridors tells the story of a hospital that is outgrowing its physical plant. (Berger)

ARE YOU EXPECTING?
RING YOUR HUSBAND TO CLASSES SPONSC
D BY, M.T.H.P. INC., AND LEARN MORE
UT PREGNANCY, CHILDBIRTH, & PARENTIN
Date: OCT. 23,
Time: 7:30
PLACE:
NEXT CLASS
WILL BE
MTN. TRAILS HLTH.

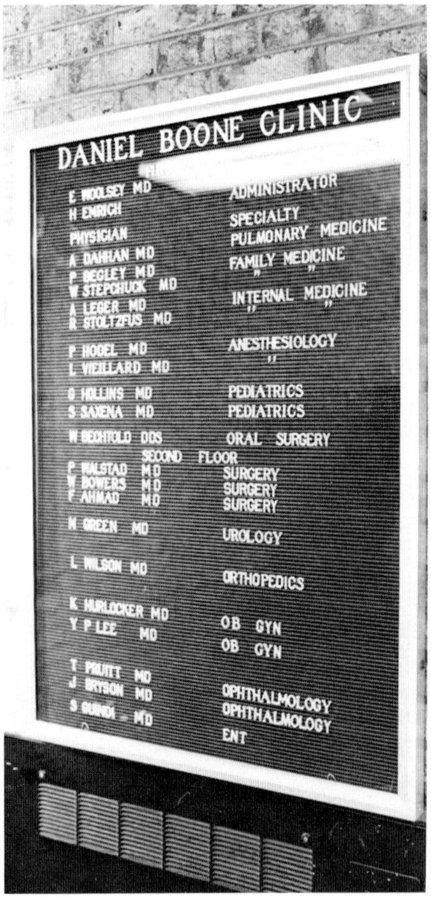

DANIEL BOONE CLINIC
E WOOLSEY MD
ADMINISTRATOR
H EMRICH
SPECIALTY
PHYSICIAN
PULMONARY MEDICINE
A DAHHAN MD
FAMILY MEDICINE
P BEGLEY MD
W STEPCHUCK MD
INTERNAL MEDICINE
A LEGER MD
R STOLTZFUS MD
P HODEL MD
ANESTHESIOLOGY
L VIEILLARD MD
G HOLLINS MD
PEDIATRICS
S SAXENA MD
PEDIATRICS
W BECHTOLD DDS
ORAL SURGERY
SECOND FLOOR
P WALSTAD MD
SURGERY
W BOWERS MD
SURGERY
F AHMAD MD
SURGERY
N GREEN MD
UROLOGY
L WILSON MD
ORTHOPEDICS
K HURLOCKER MD
OB GYN
Y P LEE MD
OB GYN
T PRUITT MD
J BRYSON MD
OPHTHALMOLOGY
OPHTHALMOLOGY
S GUINDI MD
ENT

Harlan, Kentucky

Most modern hospitals have temperature-controlled environments. The Harlan Appalachian Regional Hospital had no air conditioning until 1979. Funds totaling approximately half a million dollars were raised by hospital employees, physicians, and county residents to complete the project and, thus, provide an environment more conducive to the healing of patients.

Previously, fans were set up in wards to cope with 90-degree temperatures, and patients needing other than emergency surgery were often taken to other hospitals. Neither of these practices presented acceptable alternatives to Harlan residents or the medical community as a whole.

In 1978, William Zuercher, the hospital Administrator, stated:

Doctors are unwilling to admit patients when the weather is hot — heat aggravates the patient's condition ... The hospital [is] losing money in quantities to make a difference ... With air conditioning, we sincerely hope fewer people will find it necessary to leave the community to receive care.

The Harlan Appalachian Regional Hospital is located in Harlan, a town that is almost in the center of Harlan County. Next door to the hospital is the Daniel Boone Clinic. Two other small facilities, the Clover Fork Clinic and Lynch Medical Services, are located within 10 and 22 miles, respectively.

The hospital, which was built as one of the UMWA's original Miners' Hospitals, currently has a total staff of 40 physicians — both in consulting and active status. Of these, 27 practice at the Daniel Boone Clinic, including 3 internists; 3 family practitioners; 2 pediatricians; 3 general surgeons; 1 thoracic surgeon; 2 obstetricians; 2 ophthalmologists; 1 oral surgeon; and 1 ear, nose, and throat specialist. In addition, the clinic's staff is augmented by a neurosurgeon, an orthopedic surgeon, and orthodontists, each of whom makes regular monthly and bimonthly visits, but has a practice in another area.

The clinic leases its facilities from the hospital, and uses the hospital's pharmacy, X-ray department, and laboratory, but is otherwise an autonomous operation. It has been in existence since 1963; an average 8,000 patients per month are seen. It offers the only obstetrical services in the county, and is the only provider within a 100-mile radius for some medical specialties. Even so, patients needing neurosurgery or orthopedic surgery must travel to Lexington, 150 miles away, where the University of Kentucky Medical School and Medical Center are located.

With a population of 41,000, Harlan was assigned to AMA demographic county group three in 1977 — nonmetropolitan counties with populations of from 25,000 to 49,999 inhabitants. Fifty of the Coal Commission's sample coal counties fell into the same group.

At the time of the AMA survey, Harlan County had 39 physicians and one hospital. Of its physician population, 11 were GPs, 7 had medical specialties, 11 had surgical specialties, 4 had other specialties, and 6 had hospital-based practices. In 1970 Harlan had 43 physicians in total.

Although figures for Harlan show a physician loss of three between 1970 and 1977, the county is still well above the national average for counties of a similar size in the availability of physicians, and is close to the national average for hospital facilities. In 1977, same-sized coal counties had an average of 0.68 physicians and 4.43 hospital beds per thousand, according to AMA data. In Harlan County, the figures were 0.95 physicians and 4.37 hospital beds. Nationally, the averages were 0.78 and 4.40, respectively.

Top: The Harlan Appalachian Regional Hospital and the Daniel Boone Clinic, which adjoins the right side of the hospital. (Corn) Bottom left: Historically, prenatal care has been low on the list of available services in coal-mining communities. This poster at the Daniel Boone Clinic indicates that the situation is changing. (Corn) Bottom Right: The physician roster at the Daniel Boone Clinic shows the variety of specialists that are available through the clinic. (Corn)

Fulton County, Illinois

Fulton County has one hospital, one clinic, and a sprinkling of physicians who are in private practice. The Graham Hospital and the Coleman Clinic are located in Canton — two blocks from each other.

Although Canton has physicians, most other towns in the county do not. Lewiston, the county seat, for instance, has only one physician, and he practices only on a part-time basis.

People who wish to obtain medical care in Fulton County generally travel to Canton. Canton's Coleman Clinic has a staff of 14 physicians with specialties in family practice, thoracic surgery, orthopedic surgery, obstetrics, pediatrics, urology, internal medicine, and general surgery. The 215-bed Graham Hospital, with a physician staff of more than 30, has the full complement of beds, physicians, and equipment necessary to meet Fulton's requirements. Although some of the miners say that they go to Peoria or Springfield for major health problems, the majority say that health facilities within the county are adequate — providing the patient is able to travel.

AMA 1977 data showed that Fulton had a total of 27 physicians involved in patient care. The figure included 11 GPs, 8 surgeons, 6 specialists, and 2 physicians in hospital-based practice. The physician population served 43,200 people. Fulton's total number of physicians had increased by only one between 1970 and 1977.

Seventeen coal-producing counties in the Commission's sample have demographic populations similar to Fulton County's. Fulton fell below average in physician/population ratios, but above average in hospital bed ratios when compared with these counties, and the national sample of same-sized counties. Same-sized coal counties had a physician ratio of 0.85 and hospital bed ratio of 4.43 per thousand residents; Fulton County numbers were 0.63 and 5.46. National averages for same-sized counties were 1.0 physicians and 4.40 hospital beds per thousand.

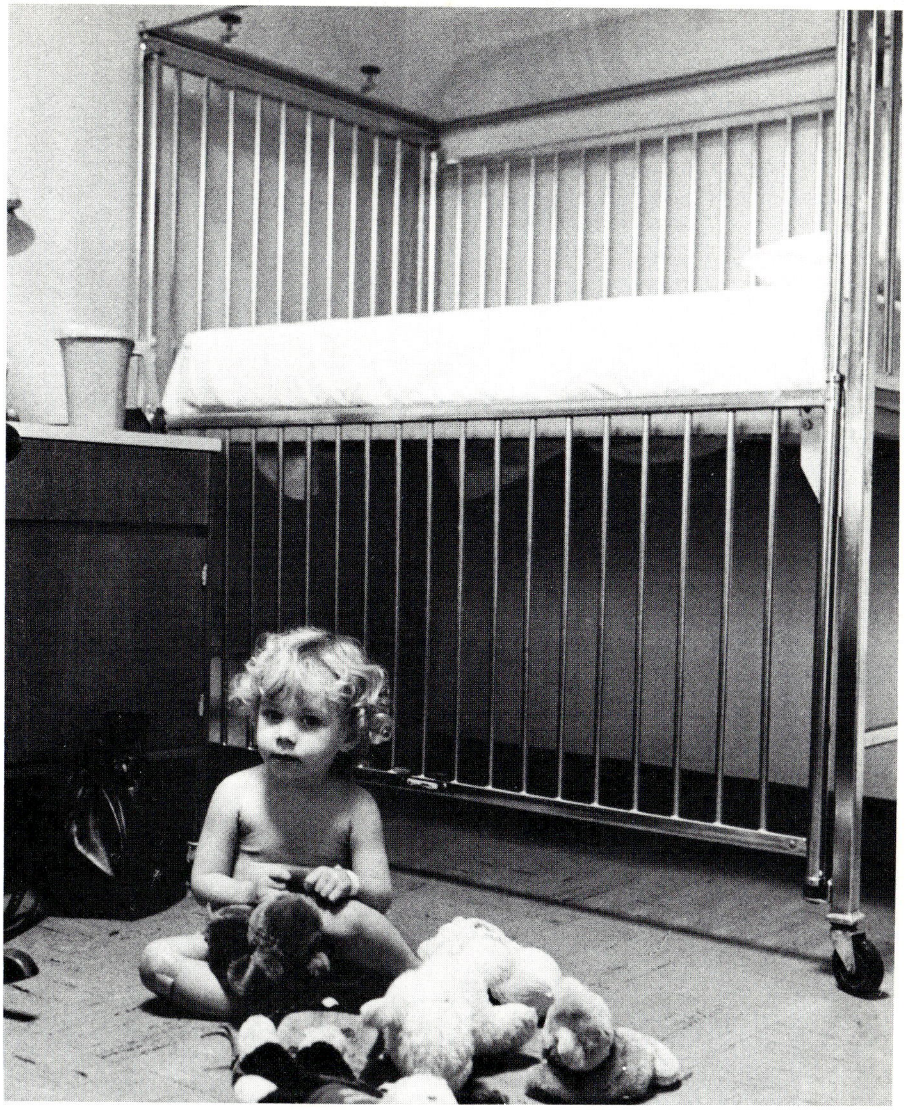

Top left: The Coleman Clinic, in Canton, Ill., has been serving its community for much of this century. Today, the clinic is outgrowing its facilities. (Wathen)

The 215-bed Graham Hospital, in Canton, Ill., provides a full complement of physicians and medical equipment to the residents of Fulton County. Top right: One of the hospital's young patients—a miner's daughter. (Wathen) Bottom: The laboratory of Graham Hospital. (Wathen)

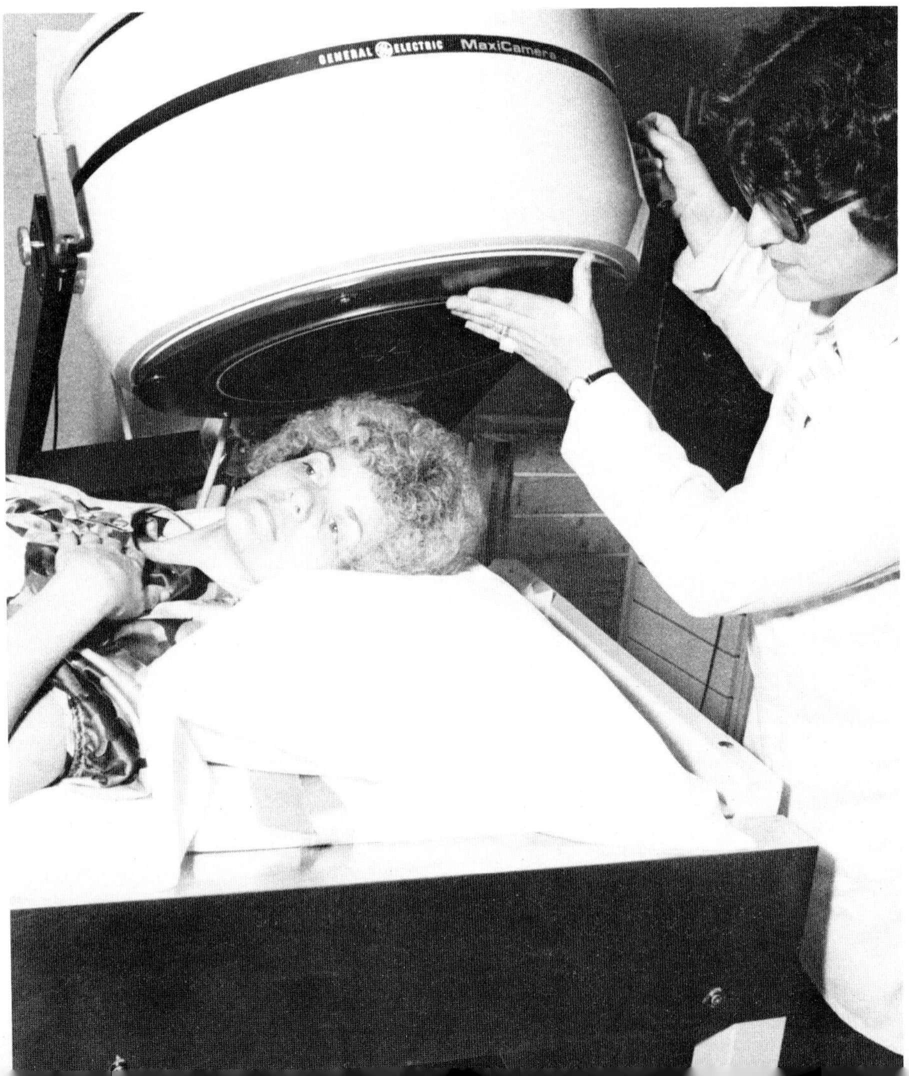
GENERAL ELECTRIC MaxiCamera

Gordon's

Walker County, Alabama

Walker County's two hospitals, with a total of 172 beds, will be closing their doors in 1980. They will be replaced by a single 227-bed facility that will have all private rooms, and nuclear medicine capabilities, and will provide staff positions for seven new physicians. The health care professionals in Walker County now eagerly await the opening of the new facility. In fact, some of the new hospital's equipment has been temporarily set up at Peoples' Hospital so that physicians and patients alike can get a taste of the capabilities that the new facility will offer.

Chester Black, the Administrator of Peoples' Hospital for some 20 years, says that the health care available in Walker County is "average." John Faust, who directs the Jackson Clinic, also located in the county, agrees, but says that care is above average for most rural communities in Alabama and will improve when the new hospital opens.

If unsatisfied with the quality of care available within the county, residents of Walker have always had the option of traveling 45 minutes to Birmingham. And, as one coal miner stated, "Birmingham is the home of the University of Alabama Medical Center, which is the most advanced medical facility in the State."

Jasper is the county's most populated area; most health care professionals, and all of the county's health care facilities are located in the town. It is estimated that the county supports a total of 35 physicians, at least 29 of whom are located in Jasper. These physicians offer specialties in family practice, internal medicine, obstetrics, pediatrics, radiology, surgery, and urology, to name a few. They provide the county with an ample diversity and supply of specialists to handle most day-to-day medical problems, but all patients needing specialized care, such as cardiac surgery or neurosurgery, must travel to Birmingham.

Walker County now has five or six more doctors than it had 2 years ago, and that number will increase again during the next year. The county was assigned to the AMA's demographic category number seven in 1977 — counties in SMSAs with 500,000 to 999,999 inhabitants. At that time Walker had a total of 30 physicians and 2 hospitals serving the county's 65,300 residents. Of those physicians, 13 were GPs, 3 had medical specialties, 7 had surgical specialties, 5 had other specialties, and 2 were in hospital-based practices. These figures highlight a trend of continuing growth in the county's medical community — in 1970, Walker had only 22 physicians.

Although it is said that the facilities and physicians available in Walker County are about average, or slightly above average, for Alabama, the county does not fare well when compared to same-sized coal-producing counties or other counties of a similar size in the 1977 AMA physician study. Five of the Commission's 169 sample coal counties were placed in the same category as Walker County. In those 5 counties, there was an average of 1.6 physicians and 5.86 hospital beds per thousand population; in Walker County the figures were 0.46 and 2.63. Nationally, the figures were 1.53 physicians, and 4.21 hospital beds per thousand.

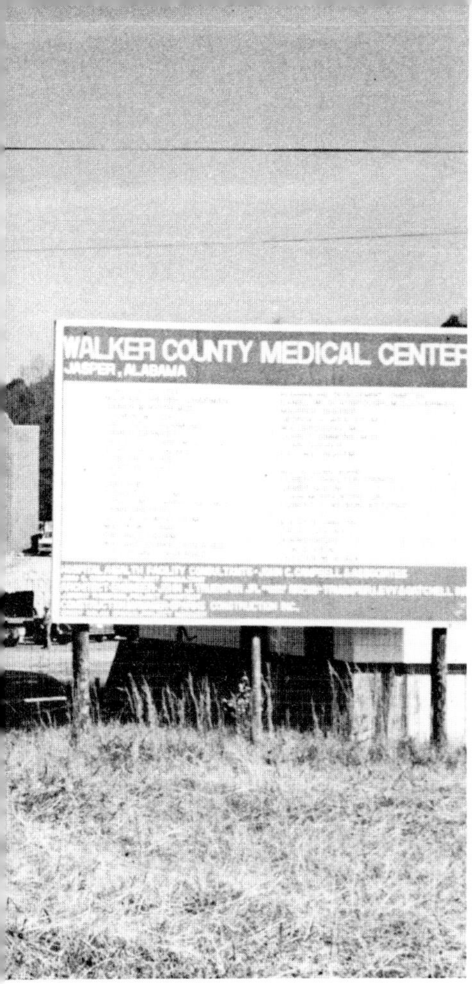

Top: In 1980 the residents of Walker County, Ala., will travel to a new hospital for medical care. The new facility will upgrade the quality of health care available. (Simmons) Bottom left: Nuclear medicine has done much to improve the ability of physicians to diagnose patients. This equipment will be installed in the new Walker County Medical Center, until that time it is being used at People's Hospital. (Simmons) Bottom right: Residents of Walker County who require delicate surgery or specialized care generally make the 1-hour trip to the Alabama Medical Center in Birmingham. (Simmons)

Economic and Community Factors Affect Health Care

The increase in the number of physicians in coal counties has been a direct result of improved facilities and intensive recruiting. During the 1950s, 1960s, and 1970s, the UMWA Health and Retirement Funds played a major role in improving the health care available to their beneficiaries — particularly in the Appalachian regions. The effort began with the construction of 10 hospitals in Kentucky, Virginia, and West Virginia. These facilities, called the Miners' Memorial Hospitals, were opened between December 1955 and May 1956. They provided 1,045 new hospital beds, brought new health care professionals to the area, and offered training residency programs in surgery, internal medicine, pediatrics, and pathology.

During the same period, the UMWA became involved in the development and financing of many health clinics, which received monthly retainers for their services.

The Miners' Hospitals and the clinics helped to improve the quality of health care locally, but the problem of finances — the problem that came to plague the health care system in the Nation as a whole — began to be felt. Many of the hospitals and clinics that depended upon the monthly retainers to subsidize their operations also depended on those funds to cover the costs of providing medical care to the many indigent residents of their communities.

Providing health care for UMWA beneficiaries and other segments of the community was more than the fund could handle financially. The hospitals became altogether too expensive to maintain and, in 1964, they were sold to the Board of National Missions of the Presbyterian Church; the hospitals became part of the Appalachian Regional Hospitals, Inc., which still operates them today.

The 1978 UMWA-BCOA contract instituted a fee-for-service payment system for health care. Thus, the 30-year-old system was eliminated. At that time, many feared that coalfield medical facilities could not survive without the fund retainers and that many of the physicians, who had been so laboriously recruited, would leave rather than wait out the uncertain future. Both predictions proved to be overly pessimistic.

To assess the impact of the new system, the ARC studied 32 hospitals and 20 clinics in July 1978. The facilities were located in eastern Kentucky, Virginia, and West Virginia. The study found that 10 hospitals reported a gain of 32 physicians, 8 re-

ported a loss of 23, and 14 reported no change. Of the 20 clinics, 3 reported physician gains, 8 reported losses, and 9 reported no change.

The ARC study asked clinic and hospital administrative staffs to describe reasons for the departures of their physicians. The following reasons were given: financial, 32 percent; personal or professional, 27 percent; desire to continue residency or education, 13 percent; retirement, 10 percent; death, 4 percent; and reason unknown, 13 percent.

The ARC study, conducted so soon after the change in financing miners' health care, was not definitive. A year later, there was no consensus among health care experts as to the net impact that the change to a third-party insurance carrier system had caused.

One of the 10 hospitals constructed with UMWA funds was located in Beckley, W. Va. Left: The Miners' Hospital in Beckley was dedicated on June 2, 1956. This photo was taken during preparations for the dedication. (UMW Journal) Right: John L. Lewis speaking at the dedication of the Beckley Hospital. (UMWA Health and Retirement Fund)

Use of Facilities

Many of the clinics and hospitals in Appalachia and the Midwest have experienced financial difficulties. Three factors have contributed to these difficulties. First, many miners who gained the freedom to choose where and from whom they were to receive medical care have taken advantage of that freedom and begun to travel outside of their communities to seek care. Second, hospitals and clinics have been forced to absorb the costs of treating the indigent poor in their communities — costs that previously had been borne largely through the UMWA retainers.

Finally, there have been substantial increases in bookkeeping costs. When the clinics and hospitals received retainers, one bill was sent to the UMWA and one check was received for payment of that bill. Today, insurance claims for each patient must be submitted, and payments of each claim recorded and deposited. If a clinic sees 1,000 miners a month, 1,000 insurance forms must be sent, and eventually 1,000 payments recorded and deposited.

The Jackson Clinic in Jasper, Ala., for instance, reports that two satellite clinics were closed because of decreased use and increased operating costs. The administrator hopes that new satellite facilities will be opened in the future. The Man (W. Va.) Appalachian Regional Hospital was forced to close in August 1978, because of the decrease in use of the hospital and, subsequently, the number of physicians in the Man area. That hospital reopened in 1979.

During that period of physician loss in Appalachia, quite the opposite situation occurred in the West. A 1979 study conducted by the Department of Energy shows that the number of physicians in the western portion of the 169 coal county sample has generally increased since 1977. Colorado lost 2 physicians, and North Dakota lost 1, but Montana gained 10, Utah gained 16, and Wyoming gained 18. Thus, the total gain has been 31 physicians. And the West continues to grow.

Recruiting and Retaining Physicians

Convincing young doctors to locate in coal-mining communities has been and continues to be a problem, especially in areas such as Appalachia. As Dr. Elwood Woolsey, Medical Director of the Daniel Boone Clinic in Harlan County, Ky., explains:

It's very difficult to recruit doctors, particularly young doctors, for areas such as Harlan or Whitesburg. We're kind of the end of the road. We get the doctors who have just gone through medical school.... They start out, a lot of them, with the idea that they'd like to go into the rural areas and do family practice medicine and then as they get further along, they get married, and their wife doesn't want to go to the tail end of creation. They want to be around a big city.

Then too, we have a problem competing with the rest of the country on the charges that can be made. Medicare and Medicaid [payment scales] are at a low level in the rural areas. If a doctor does an appendectomy in Louisville, he'll get paid more than if he does it in Harlan. We're in a position that we can't compete with what doctors are offered to go to the larger areas.

The problem of recruiting doctors was further explained by David Morrison, the Administrator of Logan (W. Va.) General Hospital. The problem he faces in attracting graduates of U.S. medical schools to his area is not income, but the deficiency of community services and opportunities in rural Appalachia. He says:

We have a very difficult time finding housing for our doctors. The rating of our schools is among the lowest in the State, and the social and recreational opportunities are lacking.

Health facilities are not a problem [the hospital was built in 1975]. We have most of the equipment that they use in medical school, and they can make more money here than they can in the big cities because the patient/doctor ratio is greater here. But historically this has just been a stopping off place. We have a lot of turnover. They come here generally because they can make more money here than someplace else. If they need to save money this is a good place.

Left: The Appalachian Regional Hospital, in Man, W. Va., is one of the original Miners' Hospitals. Because of financial difficulties, the hospital closed its doors in 1978; it has since reopened. (Corn) Right: Mining communities often face difficulties in recruiting physicians because of the lack of housing, services, and recreational facilities. (Murray)

Foreign-Trained Physicians

Foreign-trained doctors have helped to ease some of the physician shortages in coal-mining communities. They account for 4 of the 23 doctors on the staff of the Campbell County Memorial Hospital in Gillette, 24 on the 38-doctor staff of Logan General Hospital, and 10 on the 27-doctor staff of the Daniel Boone Clinic.

These doctors come from countries throughout the world and from the United States. They have one thing in common — they all receive their medical education from schools outside the United States or Canada; in addition, English often is not their native language.

Before any foreign-trained doctor can enter this country to practice medicine, he or she must be certified by the Educational Commission for Foreign Medical Graduates (ECFMG). This certification requires a review of the applicant's academic records and the passage of examinations covering medical knowledge and the English language. Passage of the examination allows the foreign graduate to enter a residency program in the United States.

Upon successful completion of that residency program, the foreign doctor must pass a second exam — the Federal Licensing Examination (FLEX) — before he or she can be licensed to practice in any State. Medical students trained in the United States take national medical boards: FLEX corresponds to these exams. Having completed all of these steps, the foreign medical graduate must obtain licensing from the State where he or she intends to practice — just like every U.S.-trained physician.

Some miners say that most of the doctors in their communities are foreign. This is not unique to mining areas. For example, in 1977, 42 percent of the 22,874 physicians providing primary care in New York City were foreign trained; 18 percent of the 4,437 practicing physicians in Pittsburgh, Pa., were foreign trained; and 22 percent of the 4,110 physicians in Houston were foreign trained.

Many of the miners complain about the foreign doctors. Their complaints are often not about the quality of care, but rather about the communication problems.

A miner from Perry County, Ill., says:

Most of the doctors around here are foreign. We've got a couple of good doctors, but I don't hear good, and they don't talk good, so I go to Belleville.

Dr. Abdul Kadar Dahhan is an internist specializing in pulmonary medicine at the Daniel Boone Clinic. He was born in Damascus, Syria; he received his medical training in Damascus and did his residency in the United States. He is not a citizen of this country, but has been practicing medicine here for the past 12 years. He speaks English with a slight accent.

Dr. Dahhan observes that there probably are communication problems with foreign doctors. He says:

I think that if a physician cannot relate to the patient, then there is no reason for them to get together. But it is very difficult to attract American graduates to this area, to rural areas in general.

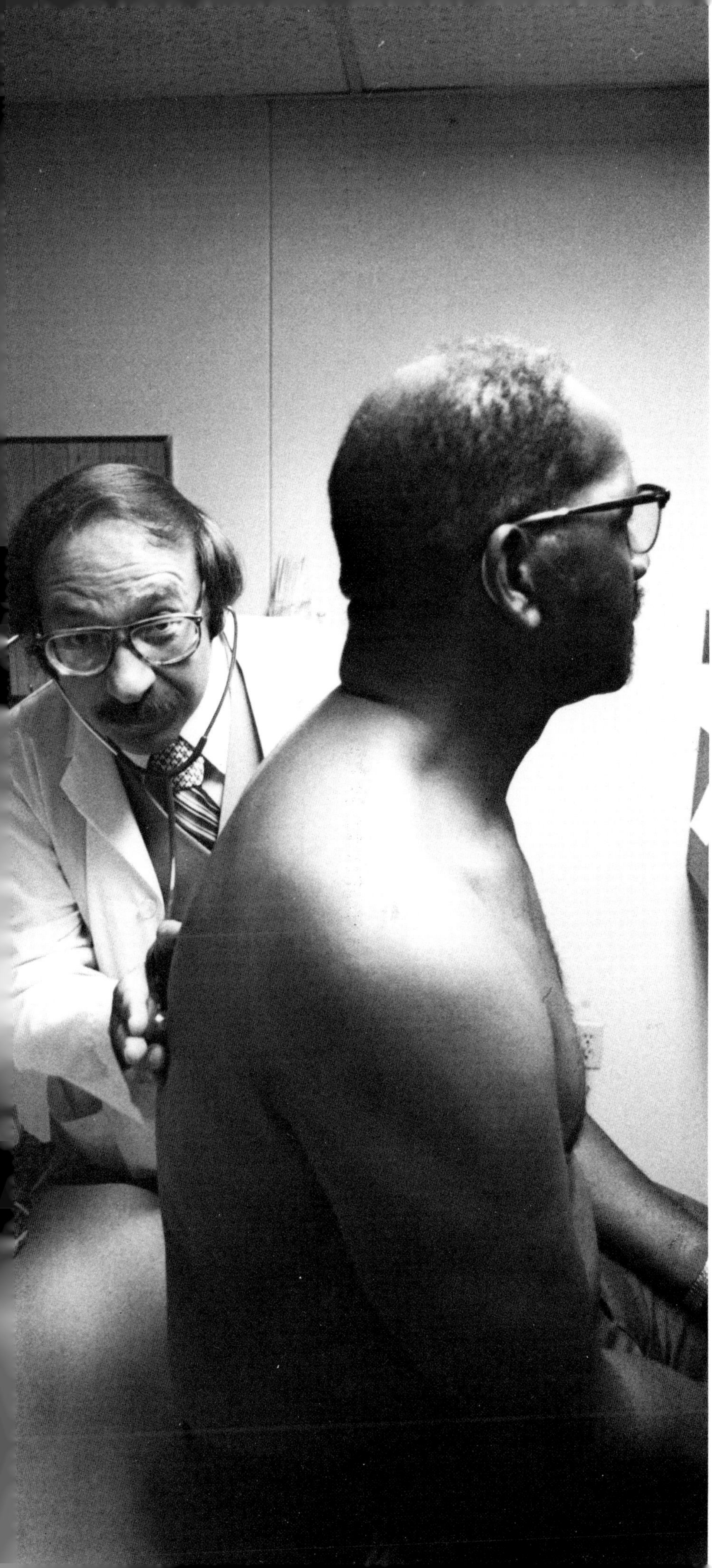

In many communities the gap in the number of physicians available has been filled by foreign doctors. Foreign doctors have become the norm at places such as (left) the Mt. Empire Clinic in Appalachia, Va., (Murray), and (right) on Jefferson Avenue in Washington, Pa. (Owre) Center: Dr. Abdul Dahhan specializes in pulmonary medicine. A native of Syria, he has practiced in this country for the past 12 years. He is a member of the staff of the Daniel Boone Clinic in Harlan, Ky. (Corn)

Federal Involvement

Although the number of physicians in coal communities has increased, there remains an acute shortage in many areas. Of the 169 counties examined, the residents of 3 of those counties are still without physicians. Furthermore, of the 169 counties in the sample, HEW rated 84 as totally medically underserved and 62 as partially underserved.

"Medically underserved" can describe a variety of different situations in a particular county. In one case, it could mean that the county physicians/population ratio is low. In a second case, it could mean that there is a high infant mortality rate—an indication that the county lacks proper prenatal, postnatal, or obstetrical services. And in a third case, it could mean that there is a high proportion of elderly or poor residents and that health care is unavailable to those people.

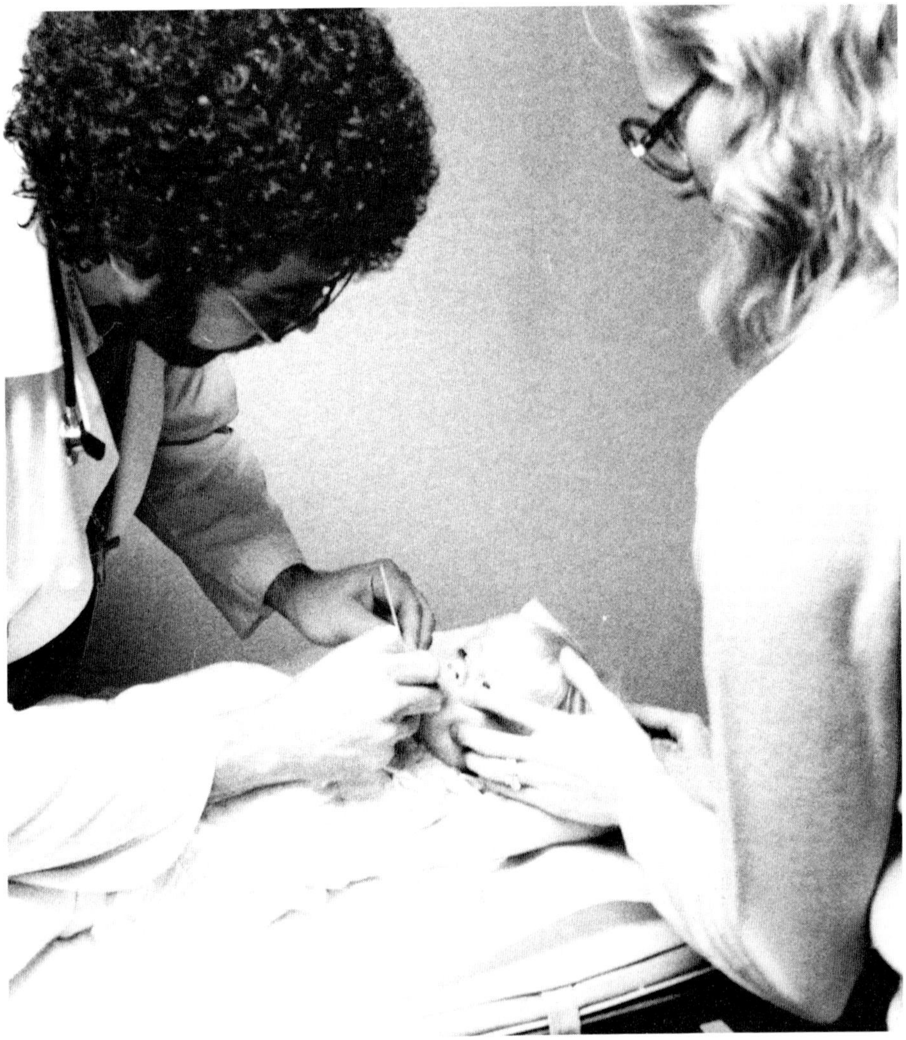

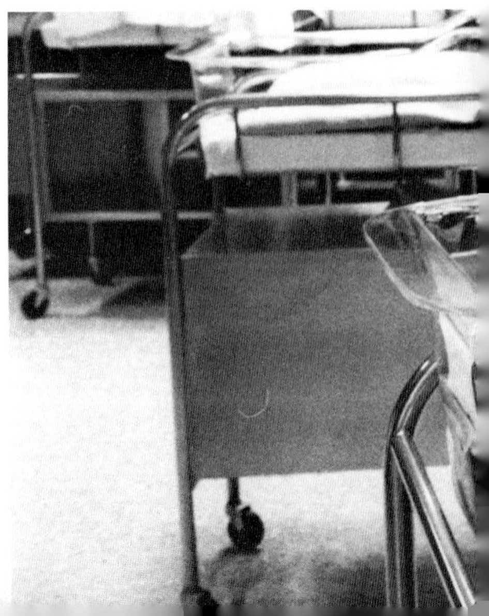

For whatever reason, 86 percent of the coal communities in the 169-county sample fit the HEW definition of a medically underserved area. Seventeen percent of the counties have high infant mortality rates; 50 percent are rated as Home Health Preference Areas — areas with a resident population that is either old or below the poverty level; and 67 percent are considered Child Health Underserved Areas.

All of these areas are considered Health Manpower Shortage Areas, and, therefore, are eligible to receive Federal assistance. This assistance can come in the form of funds and/or the assignment of personnel from the National Health Service Corps.

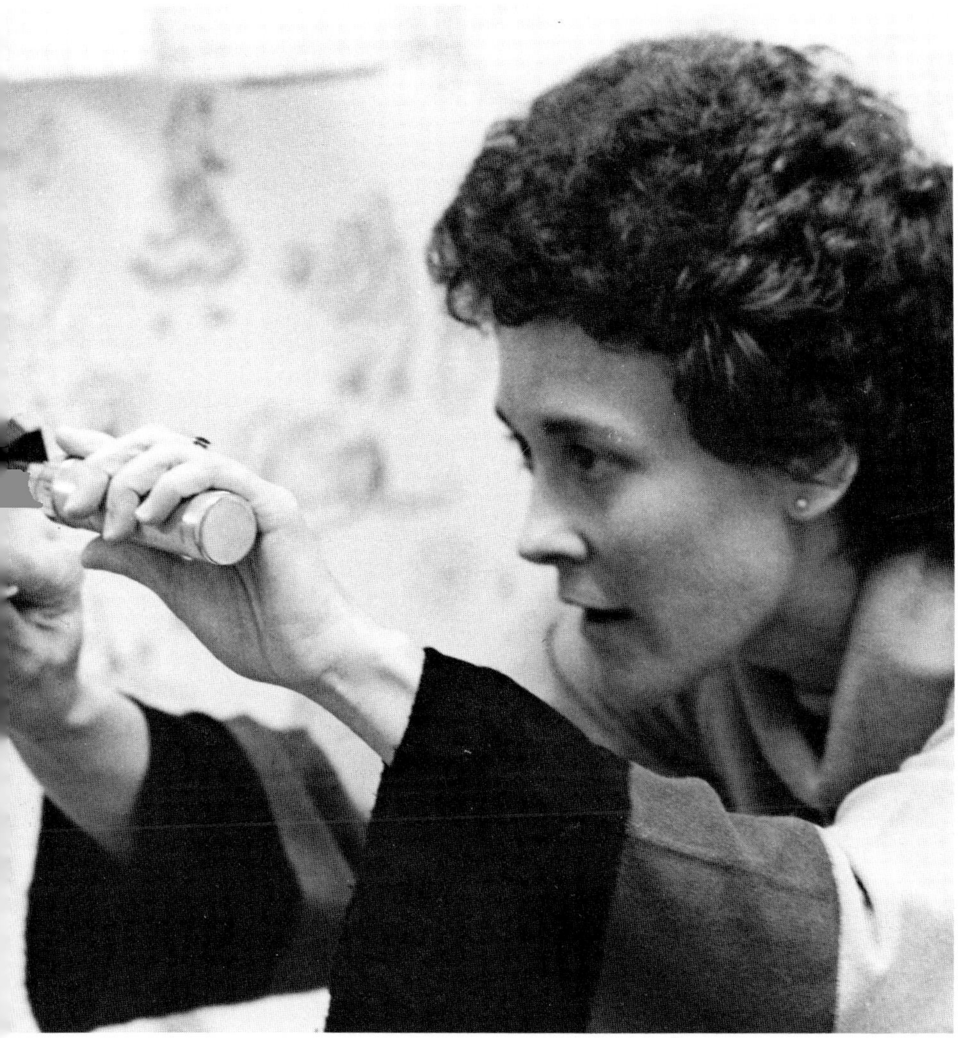

National Health Service Corps

The National Health Service Corps provides physicians, dentists, nurse practitioners, and physician assistants. These persons agree to serve in Health Manpower Shortage Areas in exchange for medical education scholarships. By June 1979, 46 physicians, 22 physician extenders, 3 registered nurses, 17 dentists, and 15 other health professionals from the Corps had been located in coal-mining communities. The majority of these persons are situated in Kentucky, Pennsylvania, Tennessee, and West Virginia.

National Health Service Corps professionals have helped fill some of the health care needs in mining communities, but the number of approved sites for Corps personnel far exceeds the number that has been assigned. There remain 30 physician openings, 12 physician extender positions, 3 registered nurse spots, 11 dentist positions, and 18 openings for other medical personnel. These positions have been approved, but the Corps has been unable to find personnel willing to fill them.

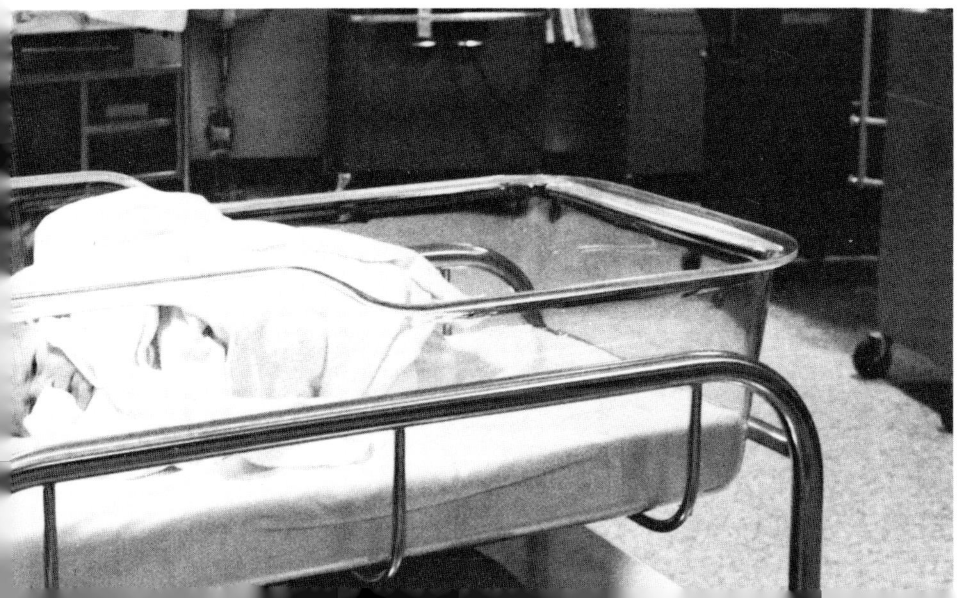

Although the Government has designated 86 percent of the 169 coal-mining counties sampled by the Commission as medically underserved, many miners and their families receive excellent care. Top left: a little girl at the Coleman Clinic in Canton, Ill. (Wathen) Top right: a widow being examined by a National Health Services nurse practitioner at the Petros Clinic in Petros, Tenn. (Corn) Bottom: an infant in the nursery of the Graham Hospital in Canton. (Wathen)

Miners' Assessment of Health Care Facilities

When asked to assess the quality of health care facilities in their communities, 55.3 percent of the respondents to the Living Conditions Survey say that they are "satisfied" and another 11 percent say they are "very satisfied."

The highest level of satisfaction is in the Midwest, where 75 percent of the miners are either satisfied or very satisfied. The lowest rating is given to the West and south-central Appalachia, where only 59 percent and 58 percent, respectively, of the respondents are either satisfied or very satisfied. Northern and north-central Appalachia receive intermediate satisfaction levels of 69 and 68 percent.

The satisfaction levels of the respondents are consistent with the availability of services, health care professionals, and medical facilities in each of the areas. Table 1 shows the percentage of services available for each area. Table 2 shows the average travel time to physicians and hospitals.

A comparison of tables 1 and 2 would seem to explain, at least in part, the reasons for the low satisfaction ranking given by respondents in the West. The West rates the lowest in percentages of available services, and also shows the highest combined travel time to doctors and hospital facilities. The northern Appalachian region, which ranks second in satisfaction levels, rates first in services available and second in proximity to doctors and hospitals. The Midwest, which ranks first in satisfaction, also ranks first in proximity to doctors and hospitals, but ranks much lower in the availability of services.

Table 1. Availability of Medical Services

Percent of Coal Miners Having Services

	Northern Appalachia	South-Central Appalachia	North-Central Appalachia	Midwest	West
Ambulance	96.6%	87.4%	90.8%	93.4%	91.0%
Drugstore	95.2	97.2	90.0	89.5	77.8
Dentist	93.9	87.1	88.2	89.9	86.9
Eye Doctor	91.7	79.6	80.3	86.0	73.8
Fire and Rescue	93.9	64.6	81.3	81.5	83.7
Paramedics	72.5	27.1	60.7	45.1	55.7
Social Service Agencies	68.1	57.5	53.1	57.3	67.4

Source: Living Conditions Survey.

Table 2. Average Minutes to Travel to Health Care Providers for Coal Miners and Their Families

Region	Travel to a Doctor	Travel to a Hospital
Northern Appalachia	23.7 Minutes	22.5 Minutes
South-Central Appalachia	23.5	30.6
North-Central Appalachia	21.9	26.8
Midwest	23.3	20.1
West	28.7	29.5
All Respondents	23.7	25.7

Source: Living Conditions Survey.

Health Status Report

Three decades ago, the health of coal miners and their families was a matter of major national concern. When the Federal Government released the Boone Report in 1947, the Nation learned of the inadequate industrial hygiene and medicine practiced in many coal mines, and of the substandard living conditions for many miners' families.

Today, health problems remain in the coalfields, as they do in the Nation as a whole. But the situation has changed dramatically, and a similar study made in 1979 would not disclose the kind of startling information contained in the 1947 report.

Taking as the baseline the year 1946 — when the Boone study was conducted — there have been substantial improvements in the overall health picture in the coalfields. A number of serious diseases, such as tuberculosis, have been reduced to the point of eradication. Other diseases, such as pneumonia, remain comparatively serious threats, with rates above the national average.

For the miner, the greatest dangers are in the mines, where he or she faces the risk of injury or death from accident, and the threat of Black Lung, a disease that is now considered largely preventable.

Industrial Health

"It gets spooky," says a miner in Harlan, Ky. "I been in the mines a long time, but sometimes I get scared. At certain times, when you get a bad top [roof], you get scared. I don't care if you've been in the mines for 40 years."

Coal mining — especially underground coal mining — ranks among the most dangerous of all occupations for fatal and nonfatal injuries. But the nature of the danger in coal mining has changed, most noticeably since passage of the Federal Coal Mine Health and Safety Act of 1969.

That Act addressed, among other issues, the principal cause of major loss of life in the coal mines — explosion.

Explosions in a coal mine occur when concentrations of coal dust or methane gas build up and are ignited by an open flame or electrical spark. Miners who survive the initial blast may suffocate as oxygen dwindles and the deadly afterdamp — an irrespirable mixture of gases, mainly carbon dioxide and nitrogen, given off by the explosion — spreads. The explosion may seal the mine and may make rescue difficult or impossible.

Explosions scarred the first 70 years of coal mining in this century with stunning regularity. Among the worst were:

- Coal Creek, Tenn., 1902: 184 men and boys dead.
- Monongah, W. Va., 1907: 360 men trapped underground and a 12-year-old trapper boy killed at the surface; the worst mine disaster in U.S. history.
- Littleton, Ala., 1911: 128 convicts-miners, mostly black, died.
- Dawson, N. Mex., 1913: more than 260 miners died in two explosions.
- Castle Gate, Utah, 1924: 172 miners died.
- Millfield, Ohio, 1930: 82 men, including high company officials on a safety inspection, died.
- Centralia, Ill., 1947: 111 miners died.
- West Frankfort, Ill., 1951: 119 miners died.
- Farmington, W. Va., 1954: 16 miners died.
- Farmington, W. Va., 1968: 78 miners died.
- Hyden, Ky., 1970: 38 miners died.

William Morrison, a 78-year-old retired miner living in Price, Utah, visits the grave of a friend killed in the 1924 Castle Gate mine explosion. His father and two brothers died in the explosion. Four other members of his family have died in mining accidents. He is a victim of Black Lung. (Corn)

Since 1970, when the Act became fully implemented, there has been only one disaster attributable to explosion. That was in Scotia, Ky., in March 1976, and involved two methane gas explosions that claimed the lives of 23 miners and three Federal inspectors.

Under current Federal law and regulations, mine operators must measure and control levels of methane and dust. Miners called "firebosses" patrol the shafts and passageways and faces of the mine, checking for levels of gas and coal dust. Preventive measures include ventilation and rock dusting, which involves spraying a white powder similar to talcum powder on the roof and ribs (walls) to control explosive coal dust. Mining equipment chewing at the face of the coal seam often now has water jets directed at the point of cutting to reduce coal dust levels.

Most importantly, equipment that is permissible in underground mining now must be explosion proof.

Fatalities

The danger of explosion has been markedly reduced, but other dangers remain, some peculiar to underground mining and some to surface mining.

The greatest danger in underground mining is from roof falls, where a section of the rock breaks away and falls to the floor. Coal falls from the collapse of sides ("ribs") also are dangerous. Mine operators observing Federal law and regulations take elaborate precautions to prevent roof falls; their miners install posts to support the roof, leave sections of coal ("pillars") to stabilize the rock overhead, and install long steel roofbolts into the roof to anchor the rock to firmer strata above.

Roof falls are the leading cause of fatalities in underground mines. The industry's success in preventing roof falls and other causes of fatal injury show up in statistics. In the years 1906—1978, there were 74,646 fatalities in bituminous coal mines. Figures for specific years are shown in Table 3.

Table 3.
Fatalities in Bituminous Coal Mining

Year	Fatalities
1906	1,581
1915	1,683
1925	1,834
1935	968
1945	925
1955	360
1965	255
1969	190
1970	255
1971	176
1972	153
1973	131
1974	131
1975	153
1976	140
1977	128
1978	103
1979*	125

**As of November 15, 1979.*
Source: Mine Safety and Health Administration, U.S. Department of Labor.

In surface mines, the rate of fatal injury is far below that in underground mines. The most serious dangers are from falls of the walls along the sides of the mine, accidents involving machinery, and slips and falls of the miner.

Fatalities per million man-hours worked (combining underground and surface mine fatalities) provides another indicator of change: 1.95 fatalities per million man-hours were recorded from 1926 to 1930; the rate dropped to 1.08 in by 1946, rose to 1.13 in 1960, and rose again to the post-World War II high of 1.37 in 1968. Since 1970, when the figure was 1.02, it had dropped steadily, reaching 0.35 in 1976 and 0.28 in 1978.

Injuries

Increasingly, coal mine injuries are coming to resemble those in the other industries. Aside from the special dangers involved in underground mining, most dangers arise from the operation of highly powered machinery, use of high-voltage electricity, and work in areas where the footing is unreliable.

Typical coal mine injuries today involve sprains and twists of the ankle or other joints,

Table 4.
Nonfatal Disabling Injuries in Coal Mining and Other Representative Industries

Lost Workdays Incidence Rates[1] Per 200,000 Hours

Year	Coal Mining	Agriculture	Contract Construction	Manufacturing	Transportation and Public Utilities
1968	8.4	—	—	—	—
1969	8.5	—	—	—	—
1970	9.0	—	—	—	—
1971	9.4	—	6.5	4.1	4.6
1972	9.3	—	5.8	4.0	4.4
1973	7.8	4.3	6.0	4.3	4.3
1974	5.7	4.2	5.8	4.5	4.7
1975	6.4	4.4	5.4	4.3	4.6
1976	7.8	4.5	5.4	4.6	4.9
1977	7.8	4.8	5.8	4.9	5.2

[1]*Incidence Rate* $= \frac{\text{Number of occurrences} \times 200{,}000}{\text{Total employee hours}}$

Source: Bureau of Labor Statistics, U.S. Department of Labor.

breaks of bones, loss of fingers, impairment of hearing, and other injuries associated also with agriculture, heavy construction, and manufacturing.

Nonfatal disabling injuries, defined as those resulting in the loss of 1 or more workdays, generally are a function of the size of the work force. They continue to be more common in underground mines than in surface mines. Representative figures for nonfatal disabling injuries per million man-hours for recent years are presented in Table 4.

Miners seem to take a philosophical view of the dangers of coal mining, even while being aware of the immediacy of those dangers.

"Sure, it's dangerous," says a miner in Red Fox, Ky. "The way I look at it, when you go down there, you never know if you're coming back out or not. I used to think about it, but I don't any more. When it comes your time, it comes your time."

In 1978, Joseph Peace, then 20, was bumped off a coal car, and a train severed his right leg. Doctors reattached his leg, and a year later he was beginning to walk. The scars on his left leg are from skin grafts. (Corn)

New Federal Laws and Regulations

When the Boone study was conducted in 1946, there was no Federal regulation of the health and safety features of coal mining.

Today, the situation is far different. But progress came slowly, and usually only in the wake of major disasters in the mines. In 1910, following the large loss of life in 1907, Congress established the Bureau of Mines as a research and fact-finding agency on coal mining. In 1941, Congress authorized Federal inspection — but not regulation — of mines in the Coal Mines Inspection and Investigation Act.

When President Truman ordered the mines seized in 1946, Secretary of the Interior Julius Krug issued a Federal Coal Mine Safety Code for use during Federal operation of the industry; but when control was returned to the operators a year later, the Safety Code became only a guideline, not a strict Federal requirement.

After 119 miners died in the West Frankfort, Ill., tragedy in 1951, however, Congress enacted the Federal Coal Mine Safety Act of 1952, which increased the inspection authority of the Bureau of Mines and empowered it to close mines involved in interstate commerce that did not conform to the Safety Code. Surface mines and small underground operations employing fewer than 15 workers were exempted.

There followed the Federal Metal and Non-Metallic Mine Safety Act of 1966, which provided the first Federal health and safety protection for noncoal miners, but none for coal miners. Then, in 1968, the second Farm-

ington, W. Va., disaster occurred. Congress moved promptly to enact the Federal Coal Mine Health and Safety Act of 1969, the most significant legislation to affect coal mining that Congress has ever passed. That Act provides the foundation for regulation of coal mining to this day. Congress amended the legislation significantly in the Federal Mine Safety and Health Amendments Act of 1977.

The Act is implemented with a set of extensive and highly detailed regulations covering virtually every aspect of coal mining. Enforcement is the responsibility of the Mine Safety and Health Administration (MSHA), which is part of the U.S. Department of Labor. MSHA has in 1979 a force of 1,027 coal mine inspectors, while — by way of comparison — the agency charged with enforcing safety regulations for all other industries, the Occupational Safety and Health Administration (OSHA), has a force of 1,564 inspectors.

Congress also authorized construction and operation of the National Mine Health and Safety Academy, located in Beckley, W. Va., to train coal, metal, and nonmetal mine inspectors and other specialists required to enforce the laws and regulations. The Academy courses are open to selected mine personnel, Federal and State employees, and students of mining.

In addition to regulation and enforcement, the Act promotes health and safety through mandatory safety training of all miners. Miners with no previous experience are required to undergo 40 hours of training before working underground or 24 hours of training before working in a surface mine. All miners must undergo 8 hours of training at least every 12 months, and they must complete task training before starting a new work assignment.

The miners themselves have always been a paramount factor in promoting safety, and the 1974 UMWA contract gave them a larger role in this regard. The contract permits any miner to withdraw, without fear of reprimand or dismissal, from any workplace that he or she considers unsafe. The contract also authorizes any miner-member of the Mine Health and Safety Committee, which every union mine has, to close the mine if the miner believes that an imminent danger exists. For a worker to have the power to close a workplace appears to be without precedent in American industry.

Signs denoting ambulance entrances are common at coal mines. This one is in Pike County, Ky. (Corn)

Aiding the Injured
Mine rescue has become highly specialized in the past few decades, with a first priority on the speedy removal of injured miners to the nearest medical facility. This was in sharp contrast to the past. A retired miner in eastern Kentucky recalls in 1979 events of more than 4 decades before:

Five years before the union, I seen my dad a-slaving. A big piece of rock hit him; broke his legs and his ribs. The company wouldn't even take him outside in a car for treatment. They said he'd have to wait for the mantrip. And the mantrip wouldn't come into the mines until quitting time. And him laying there hurt. I told those fellows, "You'll pay, I'll get even with you. Remember that!" And I did. Everyone of them got the socks whipped off of them. I give it to them. It was a debt and I paid it.

In 1979 in Gillette, Wyo., by contrast, mine rescue procedures include evacuation by helicopter to nearby medical facilities, and by aircraft to Denver, Colo., if necessary, where specialized treatment in large hospitals is available.

Mine rescue operations now draw on specially trained teams of miners who are ready to respond in surrounding mines as well as in their own. In order to keep their skills sharp, these teams compete to determine which can analyze an accident the fastest, treat the injured the quickest, and make its way out of the mine the fastest while obeying all Federal and State regulations. Miners receive no extra pay for working on the rescue team, but companies sometimes grant free vacations or offer other rewards to outstanding teams.

In Welch, W. Va., James "Hamfats" Elkins, a rescue team coach, puts his men through a practice session:

All right now, you got smoke coming out of the fan. There's several ventilation stoppages where the vent's plumb been blown out. We don't know what's beyond this point underground and we know there's at least one man out with a few bones broke. Let's hit it.

Loaded with 29-pound packs, paramedic kits, and air tanks, the team maps a proposed path into the mine and begins its simulated rescue.

Technological progress, training, and a state of readiness of rescue teams signal great change in mine rescue in the past few decades. But no perceptible change has occurred in the traditional willingness of miners to help each other in moments of peril. A wiry, gray-haired miner from Letcher County, Ky., who survived a roof fall, explains:

The bond between miners is stronger than most working people. It has to be. You let something happen in the coal mines and everybody would die together. So coal miners stick together. You can stand there and cuss a man — call him everything in the world. But if you get hurt, he's right there by your side to help you. They'll risk their lives to get you out of there. That's one thing you can count on.

But now going down the road, if you see an automobile wreck, how many people would pass that wreck? Half of them won't stop, 'cause they're afraid they'll get involved. And in a factory, nobody knows nobody.

Even the people that are as mean as hell, that don't get along with folks on the outside at all, and you don't care whether they live or die — you never hear nobody say nothing against them once they're in the mines. Miners do stick together, regardless of race, creed or whatever. I guess you could say it's a big family.

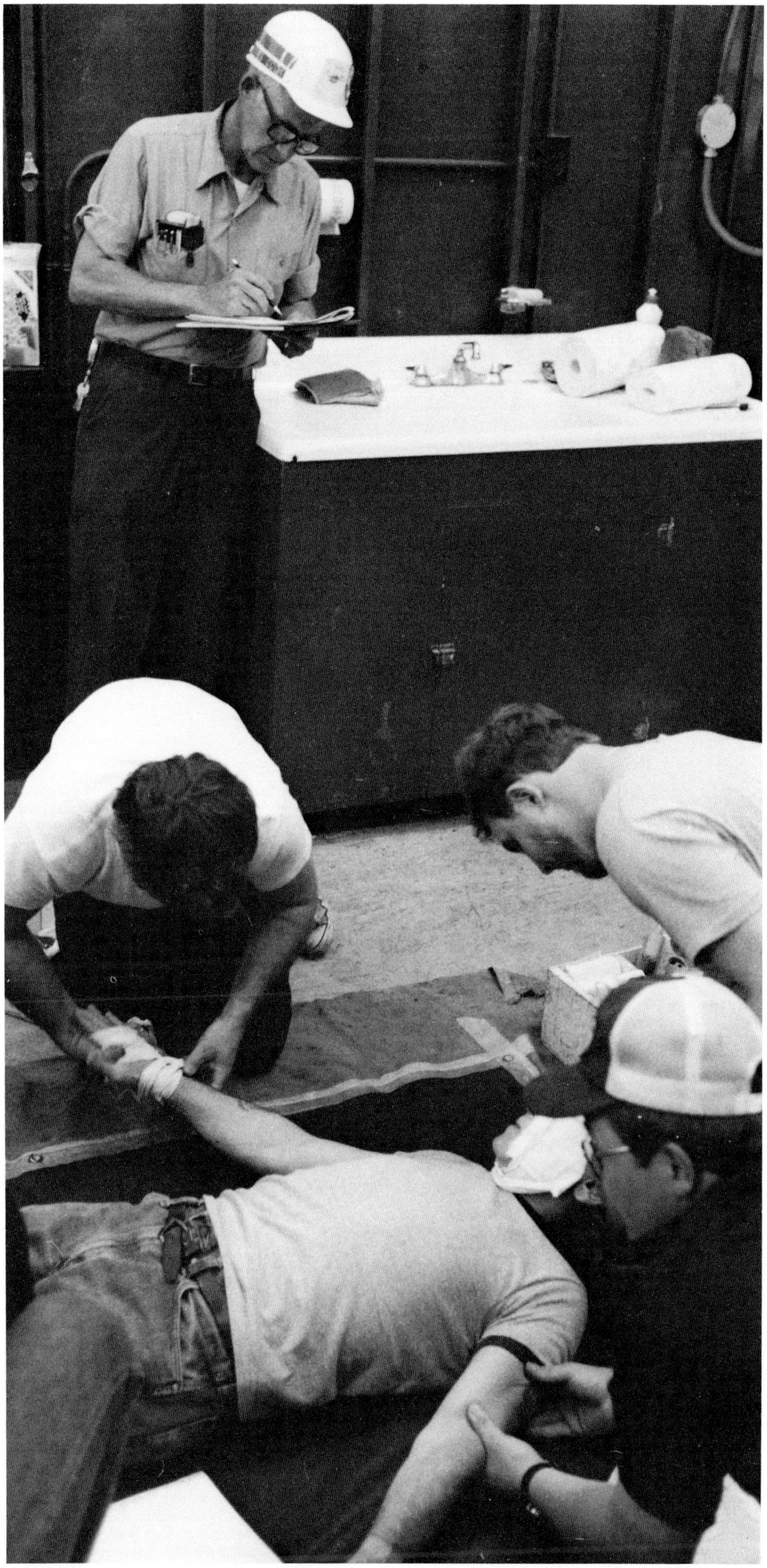

Mine rescue teams take their training and practice drills seriously. Left: Ready to enter Allied Chemical's Shannon Branch Mine, in Hemphill, W. Va., are miners Marvin Phillips, William Fink, Tennis Cook, Jr., Lewis Pierce, James Elkins, Basil King, Tom Mullins, and Tennis Cook, Sr. (Corn) Right: Allied Chemical's rescue team practices first aid on a team member alternate. (Corn)

Black Lung

Kim Sturgill, 40, of Big Stone Gap, Va., is proud to be a miner: "You've got good money, good equipment, and it's safer. The only thing I dread is the dust, that's the only thing." He dreads the dust because he has seen what it did to the older miners in his community.

In 1969, when Mr. Sturgill became a miner, statistics on the incidence of Black Lung were just becoming available, and Congress was debating legislation to protect the miner from the disease. Different studies showed that between 10 and 30 percent of all working bituminous coal miners had some form of the disease. The legislation, the Federal Coal Mine Health and Safety Act of 1969, for the first time recognized Black Lung as an occupational disease, mandated allowable dust concentration levels in the mines, and provided for compensation of Black Lung disabled miners.

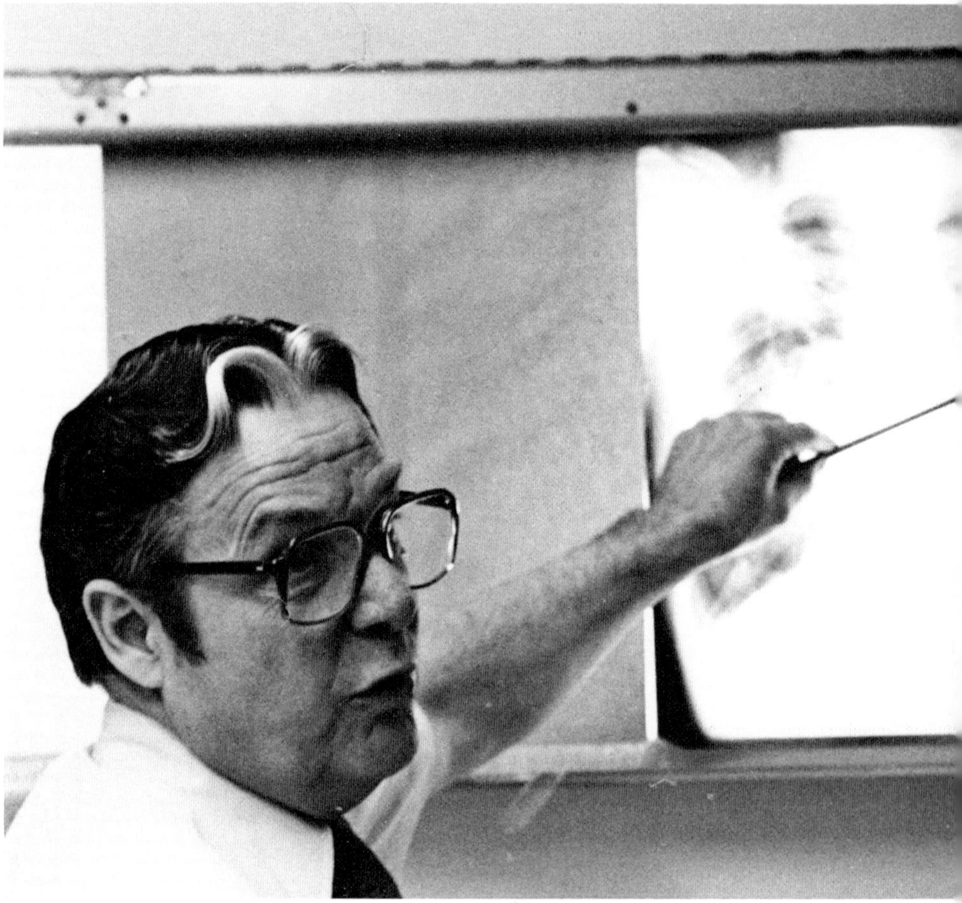

Kim Sturgill fears Black Lung, but he believes that he can take steps virtually to eliminate the chances of contracting the disease. He has an alternative that miners before him never had. He is part of the new, informed mining generation, aware that the control of mine dust should eradicate Black Lung. In his words, "I keep spraying down the sides of the walls. I don't want the Lung. I can wash the black off, but I can't wash it out of my lungs."

Contracting the Disease

Black Lung, or Coal Workers' Pneumoconiosis, is caused by inhalation and retention of coal dust. It is a disease that has plagued miners for centuries.

Every miner knows what Black Lung is. Each lives with its threat, and many older miners think that they will die from it. As one miner put it, "Coal dust ... it'll kill you. You go to town and look at some of those guys all bent over. They take a few steps and then cough a while. That's what happens to you." The symptoms — shortness of breath, coughing, loss of appetite, and difficulty in sleeping — are familiar. The pathology of the disease is less so.

There are two forms of Black Lung: simple and complicated. In the simple form, coal dust particles collect along and around the bronchial tubes — the passages that transport air to and from the lungs, and within the lungs. In the complicated form, pathological changes begin. A single mass of black tissue appears, usually in the upper portion of the lung. This may become quite large and additional masses may develop.

If the miner leaves the mines with simple pneumoconiosis, and thus discontinues the inhalation of coal dust, he or she should be able to lead a normal life. Simple pneumoconiosis may be a pro-

Ca. 1946

Top left: Black Lung causes the miner to lose the capacity to obtain and absorb oxygen. These are X-rays of the lungs of a miner in the most severe stage of the disease. (Corn)

Gonzalee Sullivan worked in the mines for 42 years. Today, at the age of 62, he is retired, disabled by Black Lung, and living in Grant Town, W. Va. Top right: Gonzalee Sullivan, with his wife and child, in 1946. (Lee) Bottom: Gonzalee Sullivan in 1979, being visited by Gary Lee Fleming, a working miner for the past 5 years. (Wathen)

gressive disease, however, if the miner continues to be exposed to coal mine dust. At current exposure levels, the likelihood of progression of the disease is small.

Complicated Black Lung is a different matter. Once the mass of black tissue has begun to develop, the disease frequently progresses until the mass, or masses, gradually constrict the flow of blood through the pulmonary vessels, first causing enlargement of the right side of the heart and later causing heart failure.

Although simple and complicated are the clinical terms used to differentiate the two forms, these are not the terms that the miners use. They refer to Black Lung by its stages: "I've got stage one Black Lung" or "He's got stage three Black Lung." These stages are the means that the Federal Government uses to specify the category of the disease for the purposes of compensation.

Black Lung is categorized by three stages. The lungs of a miner with stage one appear to be sprinkled with black pepper. The miner is considered to have the simple form of the disease. If X-rays of the lungs show black masses, about the size of a dime, stage two is diagnosed. If larger masses, ranging from the size of a quarter to the size of a 50-cent piece have developed, the miner is diagnosed as having stage three pneumoconiosis.

Miners in stage two are transitional; they may or may not progress to complicated Black Lung. Once progressed to stage three, they most certainly have the complicated form and the condition of their lungs will continue to deteriorate.

There are no firm statistics on the progression of complicated, or stage three, Black Lung, but the UMWA estimates that death or complications from the disease usually occur within a 10-year period from the time of

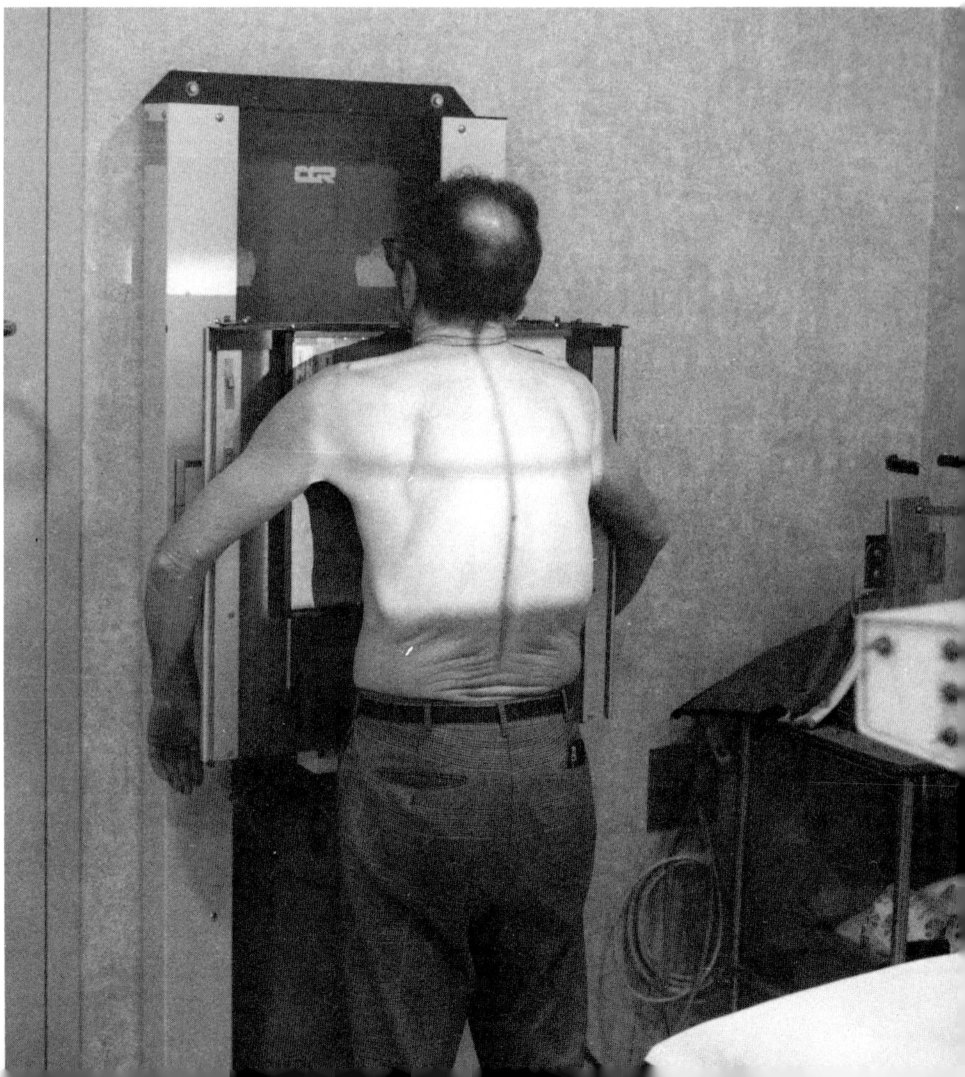

onset.

Black Lung is an elusive disease. It is impossible even to guess at the number of miners who have fallen victim to it because these people generally also experience other respiratory and cardiac difficulties. Many of them die from heart attacks, emphysema, or other respiratory problems and are, therefore, not listed as Black Lung victims—even though the actual cause of death may have been related to Black Lung.

Gad Johnson is one miner who lives with the disability that Black Lung causes. He puts on his old straw hat and starts up the steep path to his garden in Grassy Forks Hollow, Jack Horn, Ky. About every 10 steps, he pauses to catch his breath. At 64, he is disabled but undefeated by Black Lung. He considers himself "one of the lucky few" even though he has Black Lung, and has had three heart attacks, two mild strokes, and "surgery, surgery, and surgery."

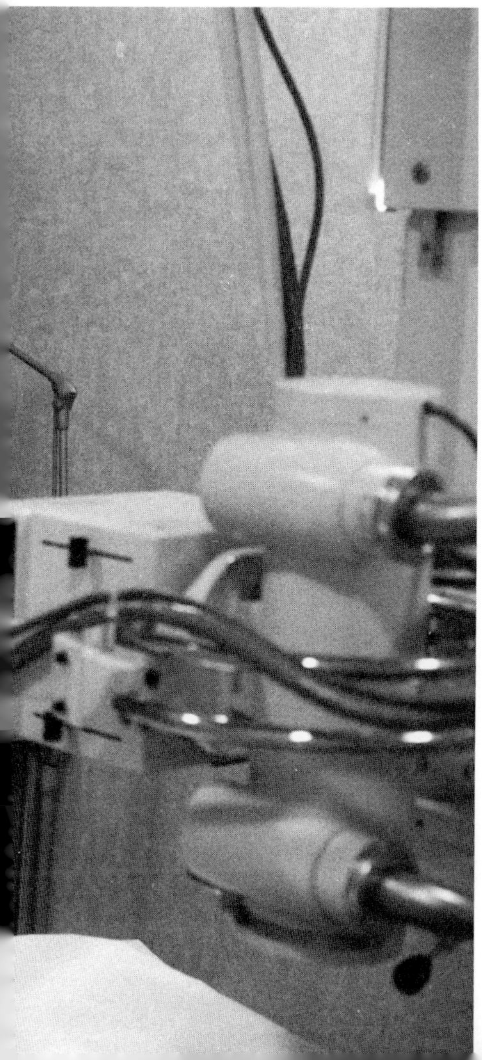

Mr. Johnson is under continual medical treatment. "You can't imagine the amount of medication I have to take. That's the only way I can live. I can't breathe. Like this morning, I got up at 2:00 o'clock and opened all the windows and doors in the house so I could get some air. I go out on the porch, sometimes in the winter when it is so cold that you almost freeze, because I'm smothering to death. I just sit there in the swing hunting for air."

Gad Johnson considers himself lucky because he is alive, and because he and his wife are able to live comfortably on his savings, Social Security checks, Black Lung benefits, and miner's pension. He has watched family members and friends die from typhoid, Black Lung, and mining accidents, and he sees neighbors who are sick and destitute. As Mr. Johnson says, "I'm one of the blessed few. But look at so many that's not and will never be."

The UMWA estimates that approximately 10 of these older miners die from Black Lung and related illnesses every day. The untimely passing of these sturdy men marks the end of coal mining when the miner was subjected to high concentrations of dust without protection. The death rate from pneumoconiosis is likely to decrease in future years because of the lowering of coal mine dust concentrations.

Understanding the Disease

Doctors and researchers have documented the existence of Black Lung in coal miners for centuries. Medical study began during the 17th and 18th centuries, and the term "melanosis of the lungs" came into general use as a descriptor for the disease.

Interest in the disease grew during the 19th century. In 1822, the term "miners' asthma" was coined because the symptoms of the disease closely resembled the symptoms of asthma, a commonly known condition. In 1837, Thomas Stratton published a paper describing the disease as it existed in the lungs of a Scotsman who had worked in the mines for 50 years. Stratton referred to the disease as "anthracosis" derived from the Greek word for charcoal. Stratton's subject had suffered from chronic bronchitis, shortness of breath, fluid in the abdomen, and swelling in the lower extremities. During the autopsy, Stratton found hard "black masses of various sizes, some as large as beans."

Top: Gad Johnson, a 64-year-old retired miner from Jackhorn, Ky., says he is happy to be alive. He has Black Lung and has had three heart attacks, two mild strokes, and repeated surgery. (Elliot)

Bottom: Raymond de Longo, from Powhatan Point, Ohio, has been a miner for 20 years. He will be 62 shortly and plans to retire from the mines. Here Mr. de Longo is having a chest X-ray—part of his Black Lung examination. (Wathen)

Over the past 150 years, others have referred to the disease as miner's consumption, silicosis, and Coal Worker's Pneumoconiosis. Even if these terms do not describe the same disease, they all describe a condition that is unique to workers who inhale coal dust and that causes the same symptoms — shortness of breath and coughing.

From the time that Stratton reported his findings until the beginning of the 20th century, it was generally accepted that coal dust was the cause of the blackened lungs of miners and the symptoms associated with that condition. Interest in the disease waned around the turn of the century, however, with reports that ventilation in the mines had improved and the disease had been virtually eliminated. Concurrently, there were other reports extolling the virtues of coal dust and its value in preventing diseases such as tuberculosis, and still others stating that only dust containing silica was dangerous.

A few investigators continued to warn of the dangers of coal dust. One was Dr. H. R. M. Landis, Director of the Clinical and Sociological Departments of the Henry Phipps Institute of the University of Pennsylvania. In an article published in 1925, Dr. Landis wrote:

A man who works in a coal mine . . . sufficiently long will inevitably develop a pneumonokoniosis. The changes may occur more quickly and may be more extensive in one person than in another . . . but there is no question as to the ultimate result. In fact, the progress of the pulmonary changes is so definite that, given the type of dust and the length of exposure, one can almost predicate the roentegeologic [X-ray] findings."

Miners continued to die from disease related to coal dust. For example, of all miners who died in 1916, cause of death due to nontubercular respiratory disease was 40.9 percent for the age group from 35-44, 84.7 percent

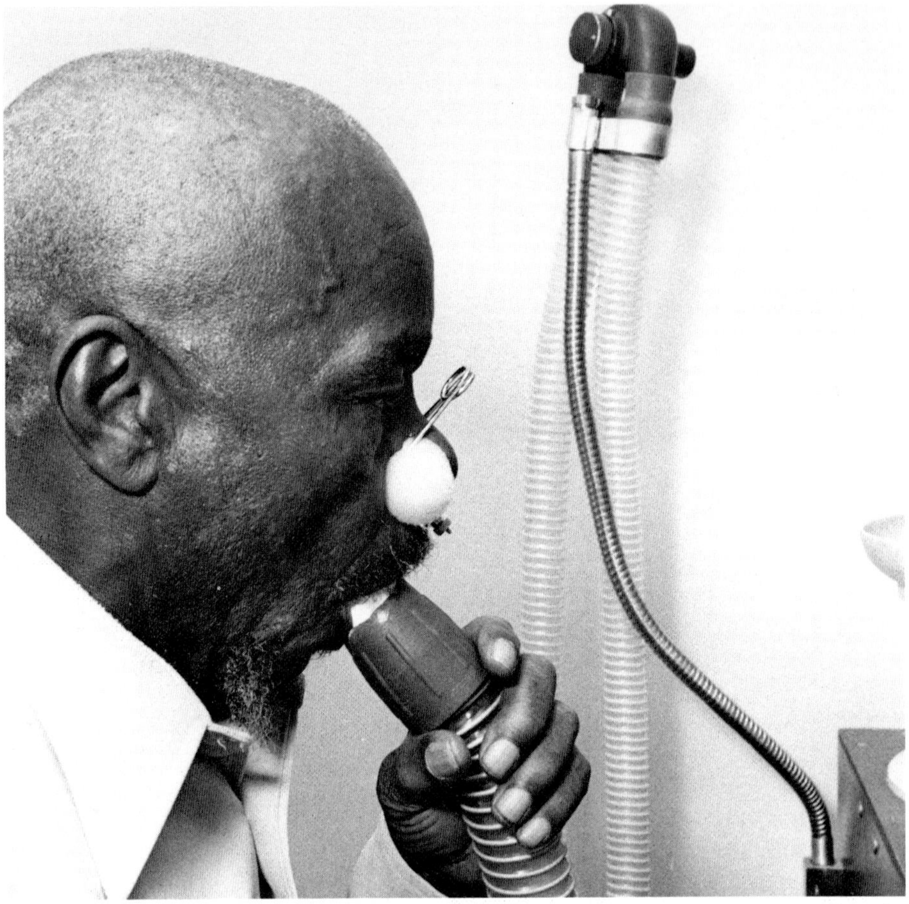

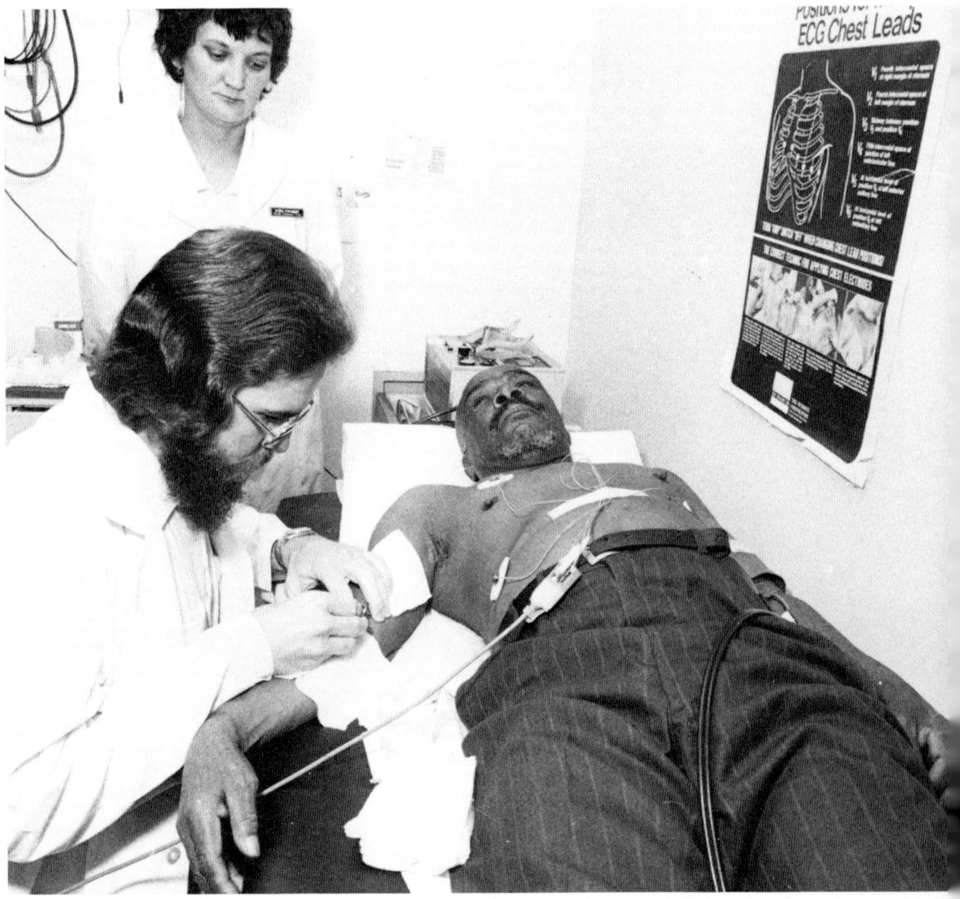

for ages 45-54, and 76.1 percent for ages 55-64.

Conditions during the 1920s and 1930s were reported to be much better, but as a miner from Logan County, Va., now disabled from Black Lung, recalls:

I go way back there to '37. They didn't have any ventilation, no sprinkler system, nothing like that in the mines. I'd come home and bathe and scrub and clean, and the next morning I'd get up and my wife would [wash] me down with a salt rag and a facial cream or something, trying to get the stuff out of my eyelids. There's no way possible that you could keep working in that and not contract Black Lung. I mean it was silly to think otherwise.

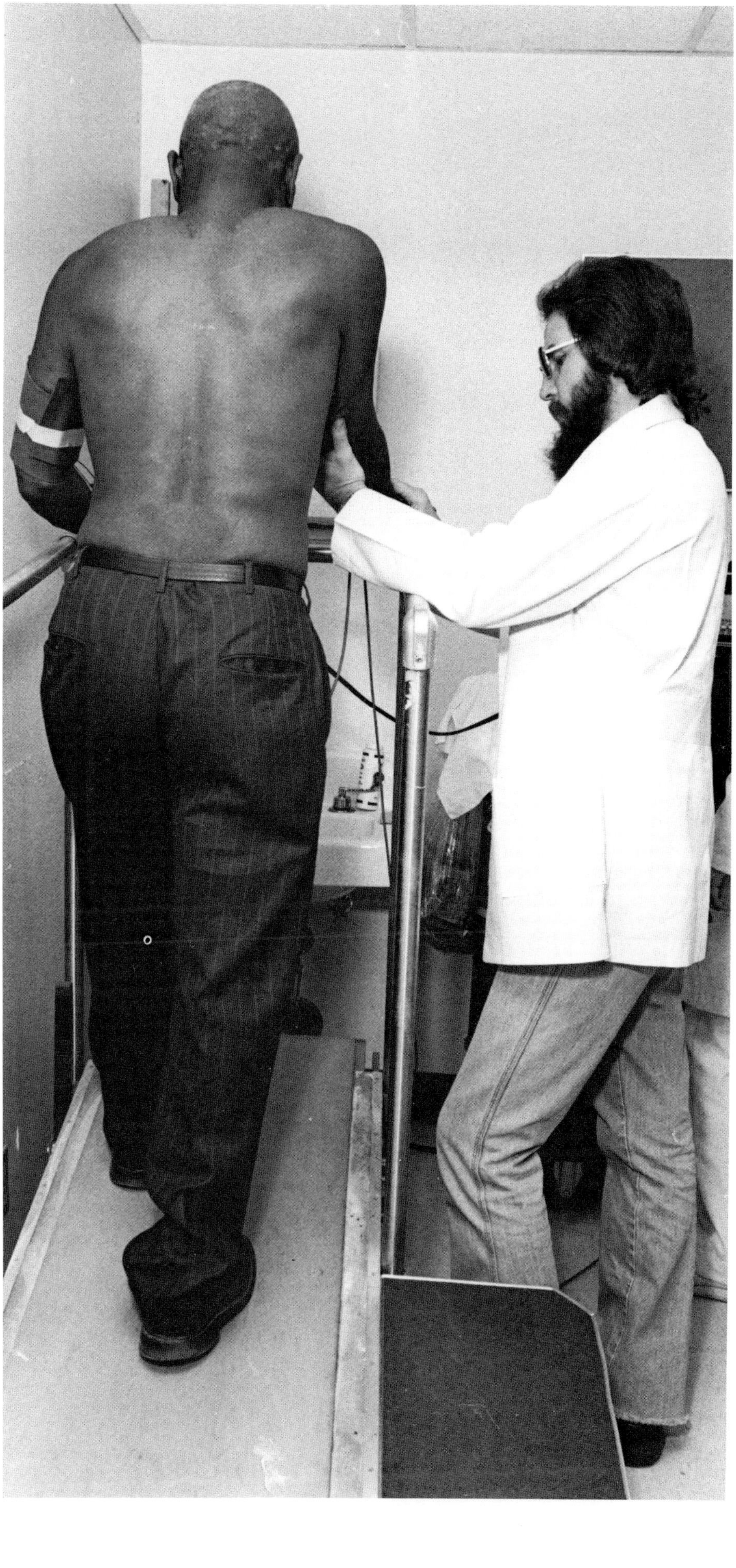

A retired miner, Clovis Casey, undergoing a Black Lung checkup at the Jackson Clinic in Jasper, Ala. Top left: Mr. Casey's lung capacity is checked. Right: The treadmill tests cardiac stress during exercise. Bottom left: Mr. Casey's blood is checked. (Simmons)

In 1937, silicosis became the first miner's lung disease to be compensated in the United States. This addition to the list of occupational diseases covered by Workmen's Compensation came on the heels of a Congressional investigation of 476 deaths that occurred during the 2-year construction of the Hawk's Nest Tunnel near Gauley Bridge, W. Va. One witness told the investigation of the death of three sons who had worked on the tunnel; each died of silicosis, and they all died within 13 months after beginning work.

It was the British, however, who finally undertook an extensive study of chronic lung diseases in coal miners. Their study began in 1937 and findings in 1942 identified a disease named "Coal Workers' Pneumokoniosis." One year after the results of this study were published, Coal Workers' Pneumokoniosis [sic] became a compensable occupational disease for the British coal miner.

The United States Government was well behind the British in recognizing the existence of the disease. A strong faction believed that Coal Workers' Pneumokoniosis did not exist in America, and that the problem for American miners was silicosis, which was already compensable. In addition, as Representative John H. Dent of Pennsylvania testified during the 1977 Congressional hearings on Black Lung benefits:

Many of the miners who have 30 years [in the mines] were never allowed to go to a doctor and have that doctor say they had any kind of a disease that was caused by the occupational disease. Anything he died from was anything but pneumoconiosis or Black Lung.

No, what he died from was pneumonia and silicosis — heart attack was the No. 1 killer disease, emphysema, things of that kind. They were not allowed to sign the death certificate that said that a man died of a disease that was contracted in the mine.

In my area, if a doctor ever once said that a man had any kind of a disease that had come out of a coal mine, he was not there the next week. He would be gone. ...

In 1952, Alabama led the way by providing compensation for Coal Workers' Pneumokoniosis [sic]. Virginia followed a few years later; that law was repealed and then reenacted in 1968. In 1965, Pennsylvania enacted similar legislation. It took another mining disaster — the explosion at Farmington, W. Va., in 1968 — to bring the problem of Black Lung to national prominence.

During the ensuing hearings on the Federal Coal Mine Health and Safety Act of 1969, one witness testified that at least 365,000 miners in the United States had died from Black Lung. At the time the figure was considered high. Only in following years, as applications for Black Lung benefits were filed, did the Federal Government become fully aware of the toll that the disease had taken on American coal miners.

Controlling the Disease

When the 1969 law was enacted, mine operators were required immediately to lower the amount of respirable dust to 3 milligrams per cubic meter of air, then to 2 milligrams by 1972. These allowable dust levels were based on experience in England, where authorities had achieved apparent success in suppressing the disease with a standard of 5 milligrams. Today, U.S. standards for dust level controls are the strictest in the world.

In addition to adhering to allowable dust levels standards, mine operators were required to make respirators available to all miners who wished to use them.

The Mine Safety and Health Administration (MSHA), the Federal agency responsible for assuring compliance with the law, officially reports that all mines are meeting the 2-milligram standard. A 1968 study showed that the average level of respirable dust in the mines was 7 milligrams per cubic meter. A 1978 study showed an average of 1.5 milligrams. MSHA is aware that many miners would dispute these figures, but believes that all miners must admit that the dust levels show a marked improvement.

Mines use water and increased ventilation to suppress the level of respirable dust. The miners themselves have respirators and surgical-type masks available to them, but often do not use them. Both of these types of devices are considered too cumbersome and ineffective. A miner in Logan County, W. Va., explains: "The respirators don't catch half the dust. They're pretty hard to breathe through too. And besides, most of the men, just like me, I chew tobacco all the time and if I wore a respirator it

would just get in my way."

Miners today are younger — about 60 percent are under the age of 35. It will, therefore, be 20 or 30 years before the effects of the 1969 2-milligram dust standard are known. Many young miners believe that they will not contract Black Lung as their fathers and grandfathers did. As one young miner put it, "I actually don't think the Black Lung will be as bad when the younger men get older in the mines, because of the way we use water and ventilation to keep the dust down."

Top: Every mine must provide each miner with a face mask. Many miners say that the face masks are cumbersome, and they do not wear them. (Murray) Bottom: Rock dusting is used to reduce coal-dust levels in mines. (North American Coal Company)

Compensating the Disabled Miner

The Federal Mine Health and Safety Act of 1969 states that all coal miners disabled by Black Lung, and their dependents, should receive compensation. The Social Security Administration (SSA) was directed to administer the program with funds provided from the General Treasury.

A retired miner who has been approved for Black Lung benefits receives monthly payments ranging from $232 with no dependents to $463.90 if he has three dependents. In addition, all medical bills for treatment of his disability are paid. For a retired union miner, the combination of Black Lung benefits, Social Security, and union pension could add up to payments of $1,100 every month. These benefits provide some pensioners with a comfortable, stable income for the first time in their lives. Since the enactment of the Black Lung Amendments of 1976, coal companies have been required to share in the financing of Black Lung payments.

Originally, in order to qualify for benefits the applicant was required to document his employment in the mine and his disability, through X-rays and other medical evidence. The SSA handled the program from 1970 to 1973, when it was assigned to the Department of Labor. Thus, miners who were approved for benefits between 1970 and 1973 continue to receive benefits from SSA; miners approved after the changeover receive their benefits from the Department of Labor. During the 3½ years that the SSA administered the program, 586,400 applications were filed. Of those applications, 415,200 were approved and 171,200 were denied.

Between July 1973 and February 1978, the Labor Department received approximately 125,000 applications; only 6,000 of those were approved. The remaining 119,000 claims were either denied or were still pending when the standards for approving claims were liberalized on March 1, 1978. The new amendments directed the Department to review all claims that had been denied or were still pending from the time when the SSA first began to administer the program. In addition, all miners receiving benefits from the SSA were given the option of submitting applications for medical treatment benefits to be disbursed by the Labor Department, whose beneficiaries were already receiving medical treatment benefits. To date, 85,000 SSA beneficiaries have applied for these benefits and 60,000 applications have been approved. Under current law, approval of these applications is automatic.

From March 28, 1979, to August 1979, the Labor Department received about 25,000 additional applications for Black Lung benefits. The Department expects that 30 to 40 percent of those applications will be approved, and that in the future the average review time for each application will be 180 days. Twelve thousand new Black Lung applications are expected each year.

The numbers of approved, denied, and pending applications are a matter of great concern to both working miners and retirees. Working miners think of their own futures when they see what has happened to members of their families and the older men in their communities, men such as the following miner.

At 62, he exhibits the symptoms of advanced-stage, complicated Black Lung; his wife says that he cannot stand up when he gets home, and sometimes needs to take a nap before he is able to eat dinner. Doctors tell the miner that he has a 70-percent disability.

He applied for Black Lung benefits in 1974, when he was unemployed. While his application was pending, he found a job in a mine in West Virginia in order to support his wife and himself. When he notified the Federal Government of his job, he was told that processing of his benefit application was terminated. His forms, medical charts, and X-rays were mailed back to him.

This is a dilemma all too familiar in the coalfields. If the miner quits work and seeks Black Lung benefits, he may or may not qualify; in either event, processing can take many months. "If I were laid off," he explains, "I wouldn't be able to get back in. They say if you're off the job for more than 30 days, you have to have a physical—and the doctors have already told me that there's no way I could pass again."

This miner's family believes that if he had gone on welfare—as many of his neighbors did—instead of seeking work, he would not be working today. He agrees: "If you were on welfare, they tried to get you off by giving you Black Lung benefits."

His family is right, at least in part. Under the law, no miner can receive Black Lung benefits while he continues to work in the mines. But there are many men like this miner who are too proud to accept welfare, who have no alternate jobs available to them, and who have families to support. So the miner reports daily for work in the mine, as fireboss, the man who inspects the mine for dangerous gas, unstable roofs, and other hazards.

A retired black lung victim from Price, Utah, with the respirator that allows him to obtain enough oxygen to function. (Corn)

Health Indicators

In 1975, the UMWA surveyed its member-beneficiaries on issues relating to health. At that time, 44 percent of the respondents perceived themselves to be in good or excellent health. In addition, 47 percent indicated that their spouses were in a similar condition, and 87 percent reported the same for their children. Of the respondents, 60 percent had undergone a physical examination within the year prior to the survey, and another 24 percent within 2 years.

The fact that 84 percent of the miners responding to the survey had undergone physicals within a 2-year period underscores their well-documented interest in and concern for their physical well being.

This section examines the practical results of that interest, in effect measuring the consequences of the succession of health care systems available to the miners since the Boone Report in 1947.

Today's miners and their families show positive signs of improvement in the overall status of their health. The miner's life expectancy — long below the norm — now approaches the norm for the U.S. population as a whole; the incidence of communicable diseases is on a par with national averages; and the incidence of infant mortality has decreased, although many mining communities continue to be rated as high infant mortality areas by the Federal Government.

One of the most effective means for looking at the quality of health care for, and at the good health of, a segment of the population is to look at the causes of death. If the causes are generally the same as for the population of the United States as a whole, then clearly the quality of care and the state of health approach the norm. If causes of death from particular diseases seem unusually high in an area, the reason may lie in the lack of the basic ingredients of health care: availability of physicians and hospitals, quality of physician care, and the other factors discussed earlier in this chapter.

Cardiovascular Diseases

In 1977, 58 percent of the deaths in the Commission's 169 coal-county sample resulted from cardiovascular system failures. More specifically, 44 percent of all deaths were attributed to diseases of the heart, and 10 percent to cerebral-vascular difficulties — strokes and the like. Cardiovascular system failures killed more people in the sample coal counties, and in the United States, than any other health problem. Only the frequency of this killer differs when coal counties are compared to the U.S. population as a whole.

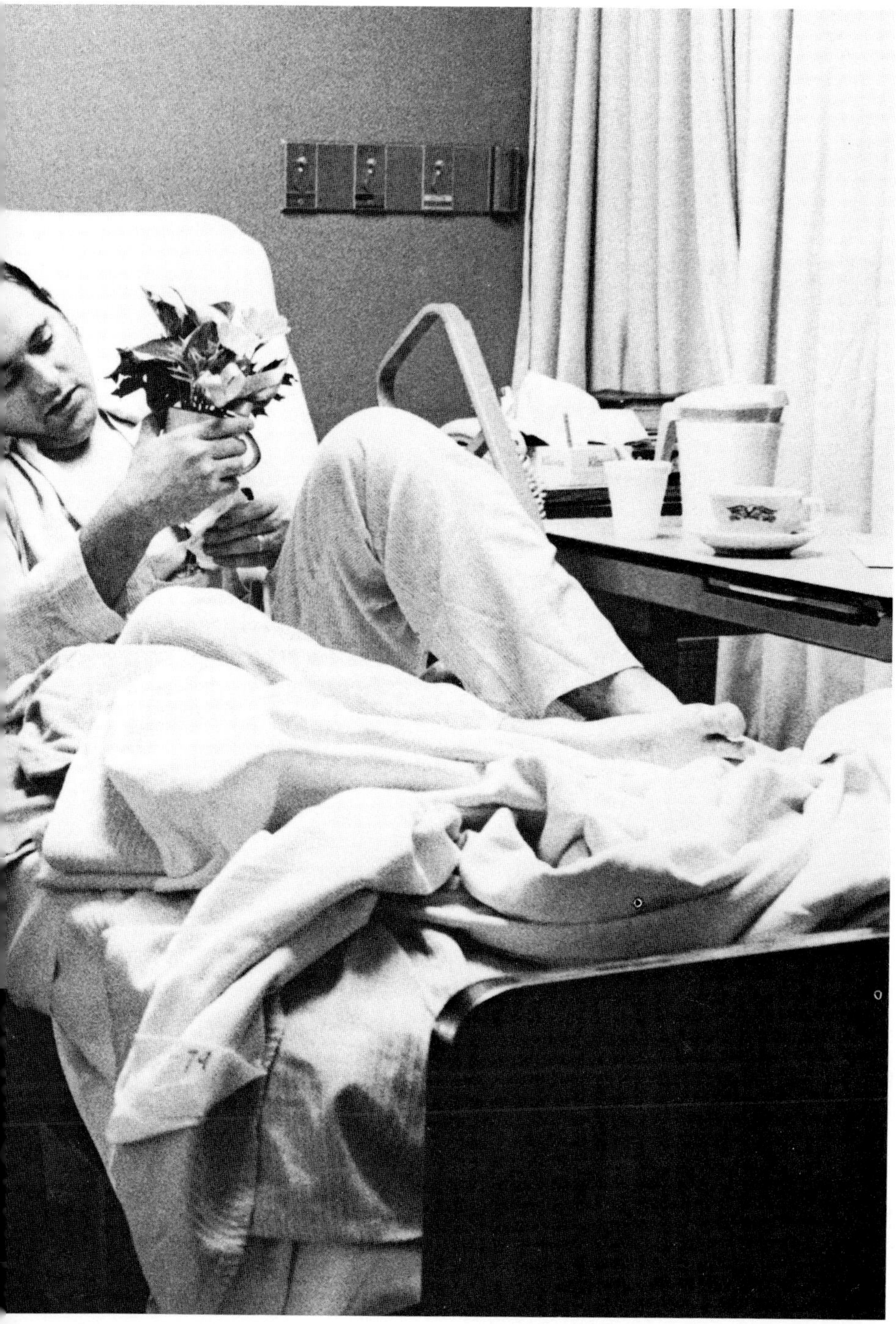

Nationally, 51 percent of the deaths in 1977 were caused by cardiovascular system failures — with 38 percent stemming from heart disease, and 10 percent from cerebral-vascular failures. The incidence of death due to these problems is more than 7 percent greater in coal-mining communities than in the Nation as a whole, and it has been consistently higher for the past 10 years.

In 1977, the National Institute of Occupational Safety and Health (NIOSH) published a report on the mortality of coal miners. As a part of the NIOSH study, the causes of death of coal miners were analyzed for a 2-year period beginning in 1969. During that period, 7,628 miners died. Of those deaths, 4,285 — or 56 percent — were the result of cardiovascular diseases.

Although the incidence of death from cardiovascular diseases is quite high in the sample coal-producing counties, there are many other diseases and conditions that coal miners and their families die from. HEW reports on a total of 34 causes of death, on a county-by-county basis, each year. These statistics provide a clearer picture of the factors that affect miners and their families.

Cancer

The number two killer in the United States is cancer, in its various forms. When taken as a whole, cancers are generally the number two killer in coal-mining communities. In 1977, 21 percent of the deaths in coal-mining counties were due to malignant neoplasms.

The prevalence of cancer as a cause of death is somewhat higher in coal counties than in the U.S. as a whole. Respiratory cancer is the more prevalent with cancer of the digestive organs and the peritoneum as a close second. This order is reversed for the U.S. population as a whole.

Tuberculosis

When the Boone survey was conducted in 1946, there was great concern about the incidence of tuberculosis. TB was a virulent killer. Today, few people die of the illness. In 1977, it was the 32nd cause of death in the United States, earning the same rating in most of the coal-mining counties in the Commission survey. Only three States — New Mexico, Tennessee, and Utah — had higher percentages of TB deaths than the coal-county average of 0.19 percent or the U.S. average of 0.16 percent.

Heart attack victim Michael Hardy, 32, is visited in his Graham Hospital room by his brother-in-law. Mr. Hardy operates a bulldozer at a large surface mine near Peoria, Ill. (Wathen)

Infective and Parasitic Diseases
The prevalence of infective and parasitic diseases is often the result of poor sanitary conditions or contaminated water. In the United States, 0.73 percent of the deaths in 1977 resulted from infective and parasitic diseases. The coal-mining county figure was 0.82 percent.

Although these percents are low, both for the U.S. population as a whole and for the coal-mining counties in general, they are much higher in the western States. In Utah, for instance, infective and parasitic diseases were the number two killer in 1977. In Wyoming, it was number 10; in Arizona, number 13; in New Mexico, number 16; and in Montana, number 18.

Influenza and Pneumonia
Influenza and pneumonia were ranked as the ninth cause of death both in the United States and the sample coal counties. The incidence of death from these illnesses is somewhat more frequent in some coal counties, however. In Oklahoma and Texas coal counties, these diseases rated fifth as causes of death, and in six other States the ranking was above the U.S. average.

Bronchitis and Emphysema
As with other illnesses affecting the lungs, bronchitis and emphysema are again slightly more prevalent in coal-mining communities. These rank as the 20th cause of death in the United States; they are the number 19th cause of death in the coal counties.

Accidents
Motor vehicle accidents accounted for 2.6 percent of the deaths in the 1977 U.S. population. They accounted for 2.76 percent of the deaths in the coal communities. Those percents make motor vehicle accidents the number 10 killer in both samples. They are, however, the number one ranking killer in Arizona coal communities, number two in Montana and New Mexico, number three in North Dakota, and number four in Virginia and Wyoming.

Other accidents account for 3.27 percent of coal-county deaths, and 2.83 percent of the deaths in the U.S. population as a whole. Accidents resulted in a high percentage of deaths in Arizona, Montana, New Mexico, Utah, and Wyoming.

Undefined Illnesses
Many of the miners interviewed by the Commission complained about the physicians and medical facilities available to them in their communities. Comments such as, "The doctor doesn't take time to look at me," "If you go to that hospital they will kill you," and "The doctors aren't any good," were frequently heard. One of the best ways to put these comments into context is to look at the number of deaths caused by undefined illnesses.

In 1977, 1.69 percent of the deaths in the United States were attributed to undefined illnesses. In the coal counties the figure was 1.98 percent. These percentages give undefined illnesses a cause-of-death rating of 14th for the United States as a whole and 13th for the coal-producing counties. Eight coal counties have ratings above the average. In Tennessee, for instance, undefined illnesses are rated as the number two cause of death. In Alabama and North Dakota the rating is fourth; in Arizona, fifth; in Missouri and New Mexico, sixth; in Utah, seventh; and in Texas, ninth.

Homicides and Suicides
Both homicides and suicides are below the national average as causes of death in coal-mining communities. Suicides rate 17th as a cause of death in the Nation. The rating is 18th in the coal counties. Most of the coal counties fall well below this figure; however, States such as Arizona, Colorado, Montana, and New Mexico raise the coal-county rating.

Homicides rated as the 21st cause of death for the U.S. population as a whole. They rated number 24th for the coal-mining counties. Although most of the coal counties fell well below the average, 10 were above it. These include counties in Alabama, Arizona, Illinois, Kentucky, Montana, New Mexico, Oklahoma, Tennessee, Texas, and Wyoming.

Mental Health
Miners and their families live with many stresses. They are concerned about fatalities and disabling injuries in the mines, Black Lung, and the instability of employment. In addition, many mining communities are located in isolated rural areas where few recreational facilities exist. They therefore lack many of the important outlets for relieving day-to-day tensions.

Even with all of these stresses, the mental health of miners and their families has never been considered a major problem. The UMWA estimates that less than 1 percent of its current health insurance claims relate to mental health problems.

(Margolis)

Form
HABITS
CHECK IN
WORK SAFE
TODAY
NO SMOKING

Transportation

Transportation is a lifeline for coal miners and their families. Miners depend on personal transportation not only to carry them to the mine, which may be at the end of a tortuous gravel road back in the hills, but also to bridge the gap between their rural communities and cities, which offer the necessities and luxuries of life.

In the coalfields, the miner must drive to work, alone or in a carpool. Unlike years ago, when many miners lived in coal company camps and walked to work, miners today are scattered throughout the region surrounding the mines. Because little or no public transportation is available, getting to work is an individual responsibility.

The family also depends on personal transportation for shopping, for visiting the doctor, for taking the children to the

nearest school or bus stop, for recreation, and for emergency purposes. There usually are volunteer ambulance services in the coalfields, but the quickest transport of an injured or sick person often is in the family car or truck.

Coal Miners Value Vehicles

Coal miners devote much attention to their transportation, investing a substantial percentage of their disposable income in vehicles, and talking frequently of the role of these vehicles in their lives.

For some miners, buying a new car or truck is gratifying. These miners earn good wages, but may find suitable land and housing unavailable in their areas. While waiting to buy a house, they buy a powerful car or fully equipped off-road vehicle as a kind of self-reward for their hard work.

Automobiles and trucks have long been commonplace in the coalfields. Miners responding to the Living Conditions Survey report that 97 percent own cars and 24 percent own two cars; by contrast, 15 percent of all rural American households do not own an automobile. The pickup truck and the four-wheel-drive vehicle capable of off-road maneuvering are evident everywhere in the coalfields; 63 percent of respondents own a truck. A miner in Washington County, Pa., explains why:

We need those big four-wheel-drive trucks around here, especially to get through the snow and ice in the winter. And even with these big trucks, we still have to spend about $200 per car per year to replace shocks, tires, wheels, and to align wheels. A small car wouldn't last one winter out here.

In addition, miners report owning an array of other vehicles: bicycles, boats, campers, dune buggies, jeeps, motorcycles, station wagons, tractors, vans, and, in one case, an auto wrecker.

One mining family in Craig, Colo., for example, owns a car, a pickup truck, a jeep, and a motorcycle.

For many Americans, moreover, ownership of an automobile has a deeper meaning — a symbol of status, freedom, virility, youth. The miners are no exception; a miner in eastern Kentucky explains why he owns two Cadillacs but drives a pickup truck to work because of bad roads:

Why should I work in the mines and the wife work all day and not have something decent to drive on the weekends if we take a notion to go somewhere? I like the large type car. It's more comfortable. With the money I make I shouldn't have to settle for something else just to get up the road.

Overleaf: Unit trains blocking traffic, coal trucks hauling on public roads, four-wheel drive vehicles, and personal cars in Norton, Va., a coal town, characterize the transportation environment of miners. (Murray)

Top left: Reminders of mining, especially coal trucks on the roads, pervade some rural areas, as in Panama, Okla. (Simmons)

Bottom left: Despite rotating shifts, back-to-back work, and overtime stints, miners sometimes carpool. These miners commute to their surface mine from Yampa, Colo. (Corn)

Top right: Cars, coal trucks, and other traffic regularly clog downtown Harlan, Ky. (Corn)

King Coal
Wild, Wonderful
381
WEST VIRGINIA

Miners own a wide variety of vehicles for work and pleasure: cars, vans, campers, off-road vehicles, ski-doos, and even airplanes. Top: Miners gather for a holiday at Kentucky Lake in western Kentucky. Bottom left: Some miners prefer the Cadillac for the family car. Bottom right: Miner Allen Grable whizzes around western Kentucky on his motorcycle. (Wathen)

Automobiles Opened the Coalfields

Coal miners have long known the importance of personal transportation. Without cars, they were tied to the rural, sometimes grimy environment of the old coal camp towns. Even if mid-century miners could not find housing away from mine sites, or if conditions of local roads discouraged traveling, personal cars gave a semblance of freedom of movement.

As early as 1922, the U.S. Coal Commission commented that more convenient transportation was creating in the coalfields "a tendency to desert some of the smaller mining patches and live in nearby cities."

Modern transportation not only increased the individual mining family's mobility but also broadened the exposure of the mining community as a whole to the larger world. The Federal Writers' Project found in 1939:

As motor highway transport has advanced, progress has been made in penetrating the eastern [coal-mining] section. Two U.S. highways now traverse the area, and a growing network of modern roads is steadily reducing its former isolation.

Accessible personal transportation contributed significantly to the demise of camp towns, thus further expanding the miner's living choices. The Boone Report explained in 1947:

The wide development of networks of all-weather roads, together with the expansion of public transportation facilities and increased ownership of privately owned vehicles, so enhanced the accessibility of once-remote areas and so lessened the traveling time from mine sites to urban centers and other communities that changes in the status of company towns became inevitable. This improvement in transportation ... created a tendency on the part of some operators to dispose of the housing they had originally constructed.

Ca. 1946

Ca. 1946

Most operators closed their camps, but others did not. A young Pennsylvanian miner, who lives today in a former company town, says that miners in the 1940s and 1950s in that area had little choice of where they lived:

Now the miner is like everyone else. We've got transportation. We can live wherever we want. We can live as close or as far away from the mine as we like.

Construction of the national interstate highway system further opened the coalfields to the freedom of personal transportation and enabled miners to extend the boundaries of their commute. The interstates came first to the coalfields of southwestern Pennsylvania and southern Illinois in the late 1950s, then to parts of Kentucky, Virginia, West Virginia, and Wyoming in the 1960s and 1970s. The interstate highways also spurred economic development, in the form of new and more diversified industry and tourism.

In 1979, conditions in the coalfields have evolved to the point that private ownership of a vehicle is practically a prerequisite to employment as a coal miner.

Personal automobiles and public transportation were beginning to open the coalfields in 1946 and to end camptown isolation. Top: Miners board buses inscribed, "Serving West Virginia's Billion Dollar Coal Fields." Bottom: Welch, W. Va., in 1946. (Lee)

Traveling to Work and Elsewhere

Miners spend more time traveling to work than do most Americans. The average one-way commuting time for miners, according to the Living Conditions Survey, is 29 minutes, compared with the nationwide average of 20 minutes in 1975.

Another study indicated that miners traveled 26 miles to work on average, which was twice as far as other rural workers (11.1 miles) and three times as far as urban workers (8.4 miles).

The longest one-way commutes for miners — 32 minutes on average — are in central Appalachia. That is followed by average commuting times of 29 minutes in the Midwest, 28 minutes in the West, and 26 minutes in northern Appalachia.

Average commuting times in the West may be somewhat misleading, because many miners live close to the mines, while some — unable to find housing — commute unusually long distances. Thus, some miners working in Craig, Colo., report daily commutes of 60 to 70 miles, a routine that they expect to maintain until they can find suitable local housing.

Carpooling

Some miners want carpooling for the trip to work or to find other methods of sharing the financial cost and the wear and tear on vehicles. Miners talk of banding together to buy a bus or van, and some suggest that coal companies could provide local transportation to the mine.

Factors influencing miners to look to alternatives to individual travel include the rising cost of gasoline and the cost of repairing vehicles that are continually subjected to the strains of long daily commutes.

Other factors operate against carpooling, however. The workforce at a typical mine in the East or Midwest is drawn from a

radius of up to 50 miles. At the end of the shift, workers disperse quickly. Miners change shifts frequently; they also "double back" or work back-to-back shifts, often on short notice, in order to earn overtime.

Given these conditions, miners say that the chances are comparatively few to find a group of workers all from the same general area, all working at the same time, and all working regularly on the same shift.

"Generally, it's hard to carpool around here where, you know, everybody lives up a different hollow, in different directions," one Logan, W. Va., miner says. "It's hard, you know, I've only been able to carpool about 5 months in the last 4 years."

Carpooling in the West is more common. Miners from a housing

development or the same neighborhood in Craig, Colo., for example, carpool for the 30-mile trip to the big surface mines outside town.

Miners share their average 29-minute commute to work with a comparatively small percentage of American workers; only about 5 percent of all workers commuted for 25 to 29 minutes, on average, in 1975. Of that group, only 5 percent were able to join a carpool and only 4 percent found public transportation.

Shopping and Other Journeys

Shopping patterns among coal miners probably differ little from those of other rural inhabitants: to buy single items or make a quick trip, they stay close to home and shop at the nearest corner market. They often take weekly trips, however, to urban fringes or shopping malls.

To shop for groceries, miners in the Midwest travel the shortest time, 14 minutes, and miners in the West travel the longest, 21 minutes, according to the Living Conditions Survey. Average travel time for grocery shopping for all miners surveyed is 17 minutes.

Miners drive farther to large shopping centers. Western miners drive an average 74 minutes, the longest time for all regions. Miners in northern Appalachia travel the shortest amount of time to get to a large shopping center — 27 minutes. All miners surveyed average 49 minutes travel time to shopping centers.

The Living Conditions Survey also indicates that miners and their families travel between 19 and 28 minutes from home to reach a doctor and between 20 and 30 minutes to reach a hospital.

Like many Americans, miners travel on vacation. Some may fly across the country to visit family or friends, but many turn to their vehicles for the trip. The Living Conditions Survey shows that about 76 percent of the respondents travel on their vacations and about 85 percent travel more than 300 miles.

"Last year we took the jeep and drove up in the mountains around Aspen," says a miner in Craig, Colo. The year before that, he and his wife rode his motorcycle to Mt. Rushmore for their vacation. In late 1979, they planned to travel to Lake Powell, Utah, where they would rent a houseboat for a week.

Many miners from eastern mining areas drive to the ocean beaches of Virginia and the Carolinas on their vacations.

Miners in the West travel an average of 74 minutes to reach large shopping centers, according to the Living Conditions Survey. This highway is near Price, Utah. (Corn)

Moving Coal Throughout the Nation

The transport of coal from mine to customer is a significant part of the industry. Coal is a bulk item and requires transportation capable of carrying heavy weight economically.

When coal production expanded in the 19th century, barges were the main form of transport. These gave way to railroads, which by 1885 were both the principal customers and transporters of coal. Rail dominates the coal haulage field today: in 1976, 431 million tons of coal were moved by rail, 90 million tons by truck, and 70 million tons by barge; 9 million tons were moved by coal slurry lines or were used at the mine. Another 80 million tons were consumed at minemouth generating plants.

Barges moved even larger quantities of coal for a portion of the journey; more than 200 million tons were moved by water all or part of the way, although only 70 million tons were moved completely by water. Waterways are more common in the East and the Midwest; western coal moving by water must first be shipped to the Great Lakes or to the Mississippi River.

Coal slurry pipelines offer an alternative to other means of transport. In a slurry pipeline, the coal is pulverized and mixed with large amounts of water. The mixture is pumped through the pipeline, and the coal is extracted at the receiving end. Presently, the 1,278-mile-long Black Mesa pipeline originating on the Navajo Reservation in Arizona is the only slurry pipeline operating. A 108-mile-long slurry pipeline exists in Ohio, but it is not operating at this time.

Coal slurry pipelines consume large amounts of water, a requirement that is especially troublesome in water-scarce areas of the West. The pipelines

also raise issues that are familiar in construction of other kinds of pipelines, where rights-of-way must be obtained and political and other interests must be accommodated over the course of many hundreds of miles.

The transport of coal usually affects miners most immediately in their local surroundings. Coal haulage trucks and other heavy traffic damage local roads, the very roads that miners drive over to work or to go shopping. In the West, 110-car unit trains often disrupt community life when they pass through small coal towns with their endless strings of coal hopper cars.

Coal is heavy and bulky and requires massive transportation systems. It is shipped mostly by barge, unit train, and truck. (Clockwise from top left: Wathen, Murray, Murray, Corn, Murray)

Miners Complain About Roads

Of all community issues measured in the Living Conditions Survey, the one evoking the most complaints is roads. Miners were asked about the roads they take to work and about road conditions around their areas, where road construction is often the most difficult and expensive.

Nationwide, 63 percent responded that they travel over "poor" or "very poor" roads to work and 79 percent answered that they are "dissatisfied" or "very dissatisfied" with road conditions in general.

Miners in every region complain about road conditions. In northern Appalachia, the largest group — 76 percent — reports that the roads they travel to work are in either poor or very poor condition. About 65 percent of the central Appalachian miners, 48 percent of the midwestern miners, and 60 percent of the western miners respond that the condition of their roads to work is poor or very poor.

Miners in all regions are also displeased with road conditions generally — to work or otherwise. Miners in the West report the highest level of dissatisfaction: more than 81 percent said they were dissatisfied or very dissatisfied with their roads. About 80 percent of southern Appalachian miners, 77 percent of northern Appalachian miners, and 69 percent of midwestern miners respond that they are dissatisfied or very dissatisfied with roads.

Nationally, 68 percent believe that their roads are worse than most roads outside their regions.

Miners responding to the Living Conditions Survey overwhelmingly agreed that poor road conditions are a persistent problem. Far left: A road twists along Appalachian mountainsides. (Murray) Left: A crumbling road near Julian, W. Va. (Murray) Top right center: Bridges cannot readily be strengthened or widened; a bridge leading to Hemphill, W. Va. (Corn) Top far right: Sign on a road in northern West Virginia. (Wathen) Bottom right: Spring rains flood a road in Letcher County, Ky. (Margolis)

NANCY'S
Style

Sad Sam
GMC

LITTLE
Rooster
CONSTRUCKTOR
Autocar
RING

Carrying loads up to 60 tons, coal trucks batter roads, throw up dust, and intimidate other traffic. (Top left and right: Wathen; other photos: Murray)

Battered Roads Plague Some Coalfields

In some eastern coalfields, many secondary roads are products of poorly financed upgradings of former dirt roads. The roadbed, drainage, and bridges were built decades ago for the lighter vehicles of the day. Some of these roads buckle and crumble under today's heavier traffic.

In Pennsylvania, for example, a highway official points out that many rural roads were built in the 1930s under the Works Progress Administration program for farm-to-market traffic, and that the roadbed consists of a thin 8 inches of concrete covered with blacktop. At least a 12-inch concrete base is needed to support the weight of large trucks.

In much of central Appalachia, winding, narrow roads cut across rugged mountainsides and snake through tight valleys. Miners and their families share these narrow roads with school buses, commercial vehicles, and coal trucks. The steady pounding damages even fairly well-constructed roads.

"The roads will shake your car all to pieces," says one southwestern Virginia miner. "They paved that road I guess 7 or 8 years ago, the only time I can remember since I grew up. Since then, they patched it and that's it. We have the worst roads of anybody in the country."

Construction defects, such as weak shoulders, poor drainage, and recurrent potholes, often appear in roads in coal-mining areas. In parts of Appalachia, "orphan roads," built originally by mining and lumber companies and not regularly maintained as public roads, crisscross coal communities. Freezing, thawing, and flooding contribute to the damage.

Many bridges in rural areas are old and are not built to withstand heavy coal truck traffic. Although local roads may be repaired and resurfaced, bridges cannot be strengthened so readily. The Appalachian Regional Commission reported in 1974 that up to 1,100 bridges in that region were inadequate to withstand current traffic.

Top left: Kentucky officials estimate that repairs to roads that regularly carry coal trucks cost at least three times more than those for other roads. (Margolis)

Bottom left: Bob Wimsatt, a miner in Evansville, Ind., tunes his car engine. Miners everywhere are concerned about fuel economy and the price of gasoline. (Simmons)

Right: Miners and their families in the West, like this woman and child in Oak Creek, Colo., must contend with unpaved roads that have no sidewalks. (Corn)

Coal Trucks Punish Roads

The steady traffic of coal-haul trucks punishes roads in a number of coalfields, mostly in Appalachia and to a lesser extent in the Midwest. About three-fourths of all coal mined in the country at some point in its progress from mine to consumer is carried by road. Most of this movement is in Appalachia, where half the coal is trucked from mine to tipple.

Coal trucks, usually owned by small contract operators, rumble over narrow country roads, through city centers and tiny communities. One Appalachian city official reports: "There's 30-ton coal trucks that go down Park Avenue, the main street, every 60 seconds, 24 hours a day."

Roads outside towns take a beating. A miner near Norton, Va., says:

We've only got roads that's for light traffic, and we got coal trucks. The only industry we got. You take a road that's in fairly good shape, run coal trucks on it 1 month, and the bottoms are out. Anybody that's familiar with this part of the country can tell you which way the coal traffic is going, and never see a coal truck because the side of the road where the loaded traffic is going is torn to pieces and the other side is a whole lot better.

In Appalachia, coal moves over approximately 14,300 miles of road in an eight-State region. Often, coal trucks weighing 25 to 40 tons travel over roads with 15-ton limits.

A 1977 study comparing replacement and maintenance costs of roads in coal-producing Virginia counties with those in nearby noncoal counties, showed a marked difference. The cost per mile for maintaining primary coal-haul roads was $6,439; the cost per mile for other roads was $1,886.

In Kentucky, the Nation's leading coal-producing State, the situation illustrates the problems involved in coping with the road

damage. The Federal Government rates 96 percent of State-maintained roads and all country roads used for coal hauling as "poor" or "very poor" by Federal inspection standards. These roads cannot handle any increase in truck traffic, and are a constraint on miners, local residents, and coal truck operators.

Kentucky authorities estimated the cost of upgrading 1 mile of coal-haul roads to support trucks weighing 80,000 pounds (the legal limit for primary roads) at $270,000, which is three times the estimate for other road renewal.

Midwestern roads also suffer. One miner who lives near Madisonville, Ky., says: "The weather and the coal trucks is what is tearing our roads up—it's pothole city." A 1978 U.S. Department of Transportation study of coal haulage reported that coal trucks in Indiana and Illinois carried an average 60 tons and traveled more than 195,434 miles annually. In Pennsylvania, 5,102 miles of road are used daily by 10 or more coal trucks.

Pulverized roads are the most obvious result of steady coal-truck traffic. Miners also complain about the dust, noise, spillage, and sheer size of the trucks, which sometimes edge school buses and other vehicles off roads.

Hauling Coal in the West

Every day, Gillette, Wyo., is cut in two by "unit" trains carrying coal from Wyoming and Montana. These are called unit trains because they carry only coal—no flatcars, no boxcars, only open coal hopper cars. They carry coal to the big coal-fired generating plants in Texas and the Midwest.

Unit trains are usually 110 cars—or about 1 mile—in length. A typical train requires up to 30 minutes to pass through Gillette. From 6 to 20 trains a day rumble down the tracks, whistling and throwing up dust. Each train divides the northern half of the city from the southern half, stopping all traffic and emergency services. Complaints to the city council are chronic.

The situation in Gillette is not unusual in the West. All towns on the railway line share the same fate, except where unit trains can be diverted to other routes. Most of these small communities lack railroad underpasses or overpasses because of the cost required to construct them, a cost averaging about $2 million for a standard underpass or overpass.

Left: The Appalachian Regional Commission reported in 1974 that as many as 1,100 bridges in the area were inadequate. This bridge is near Pikeville, Ky. (Margolis)

Right: Bad roads mean costly repairs—an additional $200 a year for new shock absorbers, tires, alignment, and other repairs, some miners say. Miner Garry Dawson works on his car in Mead, W. Va. (Wathen)

Driving in the West

Ed Sopher, a dragline operator at a Montana surface mine, is accustomed to driving long distances. When he first moved from Illinois, he lived in Colstrip, where the town, the coal-fired power plants, and mining operations are all within view. Mr. Sopher drove 11 miles daily to his job at the Big Sky Mine. Other miners there, however, lived in Forsyth, 31 miles away.

"I didn't want to have to drive 2 hours every day back and forth to live in Forsyth," he explains. "A lot of guys do it, though."

Ed Sopher still must drive over great expanses of prairie for basic services. "It used to be we'd have to go 90 miles round trip just for groceries when we lived in Forsyth. Of course, we don't have everything now, but it's a start. Tomorrow, we'll go to Billings, 130 miles one way, for things I need for the house. We go there once a month. Mary keeps a list of the things we have to have."

The Sophers' experience is shared by other western miners responding to the Living Conditions Survey, who report they travel an average of 73 minutes to reach a large shopping center.

Definitions of long distances remain a subjective matter in the West. One family sees matters its own way. Leonard and Sue Colbin — he is a heavy equipment mechanic at Big Sky Mine — prefer to live in Forsyth instead of Colstrip. "You are more isolated there," Mrs. Colbin says. "Here, we've at least got passenger trains and buses, so you can get out without having a car. That's not possible in Colstrip. You need a car to get anywhere."

Regardless of proximity to work or shopping, many western miners deplore road conditions. In the Living Conditions Survey, more than 80 percent of them report that they are dissatisfied or very dissatisfied with local roads. Approximately 71 percent

say that the condition of roads traveled to work is poor or very poor.

A miner near Price, Utah, says:

We have the worst roads in the country. In the wintertime, especially, the roads get torn up. They never fix them. And all winter you see people stuck all over the highway.

Severance Taxes Pave Some Roads

Many miners believe that their roads should be maintained or improved with funds from coal severance taxes levied on the amount of coal mined or on the value of coal produced.

Severance taxes on coal, oil, and other minerals have been imposed by local governments since 1846, and today at least 14 States as well as individual counties tax the coal extracted from their land. Miners argue that this money should be returned to the region from which it was taken to repair the damage caused by loaded coal trucks.

Use of tax money always raises debate, and miners are outspoken. A surface miner in Colorado says:

You can't imagine the road conditions — they're awful. The major coal mines are located in Gunnison County and are paying taxes to Gunnison. But I haven't seen any coming back into Delta County which is where most of the miners live and where most of the impact is being felt.

Although a number of States impose coal severance taxes, most use only a small portion of the revenues to improve roads. Many State legislatures deposit coal tax funds into the general fund and allocate expenditures administratively.

Montana, for instance, charges a 30-percent tax on the value of coal at the mine mouth. Until July 1, 1979, the State allocated 9.75 percent of the tax revenues to the Coal Area Highway Improvement Account and the remainder to other accounts, including a Coal Tax Trust Fund. The trust fund is intended to reimburse future generations for extracting the State's nonrenewable resource in coal. On July 1, 1979, the State discontinued the special allocation to the highway improvement account and doubled the contribution — from 25 to 50 percent of all coal severance tax revenues — to the trust fund.

Some States have no formula for distributing severance tax dollars and make discretionary use of them to finance a program or an area. Arizona spends most of its coal taxes for education; Colorado distributes its coal severance revenues back to coal-producing counties. Ohio uses coal severance taxes largely for reclaiming land following surface mining operations. Although Virginia presently does not impose a coal severance tax, individual counties do, and they use this money for coal-road maintenance.

Potholes and pickup trucks, such as these in Hemphill, W. Va., are familiar sights for many miners. (Corn)

Lifestyles

Coal miners and their families are no longer isolated from the mainstream of American life.

"They want the Good Life," observes a bank president in Appalachia. "When the miners have their vacations, you think it's a parade down here. You see boats being pulled by cars, often Cadillacs, kids hanging out of the windows."

The work that provides for those amenities is unusual. The underground miner works in dark tunnels, sometimes less than 3 feet high, a work world that brings a commonality of feeling to those who share it, no matter where they live above ground. Surface miners are a somewhat different breed, more akin to construction workers; they, too, have camaraderie at work and in leisure hours.

Although more first-generation miners are entering the work force, it is still common for many young workers to follow a family tradition. "My father was a coal miner, his sons were coal miners, and my son tried to be a coal miner," says a miner born and raised in the east Kentucky mountains. Approximately 74 percent of the miners interviewed in south-central Appalachia say that their fathers were coal miners. In north-central Appalachia, the figure was 56 percent, in the Midwest it was 44 percent, and in the West 24 percent.

In Appalachia, tradition, ex-

Overleaf: Miner Willard Tharp, his wife Jeanette, and sons Leland, Keith, Jeffrey, and Devin, of Orangeville, Utah. (Corn)

Miners are no longer isolated from the mainstream of American life. Increased wages and shorter hours allow them to enjoy lifestyles unknown to previous mining generations. Left: Kentucky miner Douglas Joe Harris and his family relax on their outboard. (Wathen) Right: Miner Bill Robinson and his daughter at their home in Gillette, Wyo. (Murray)

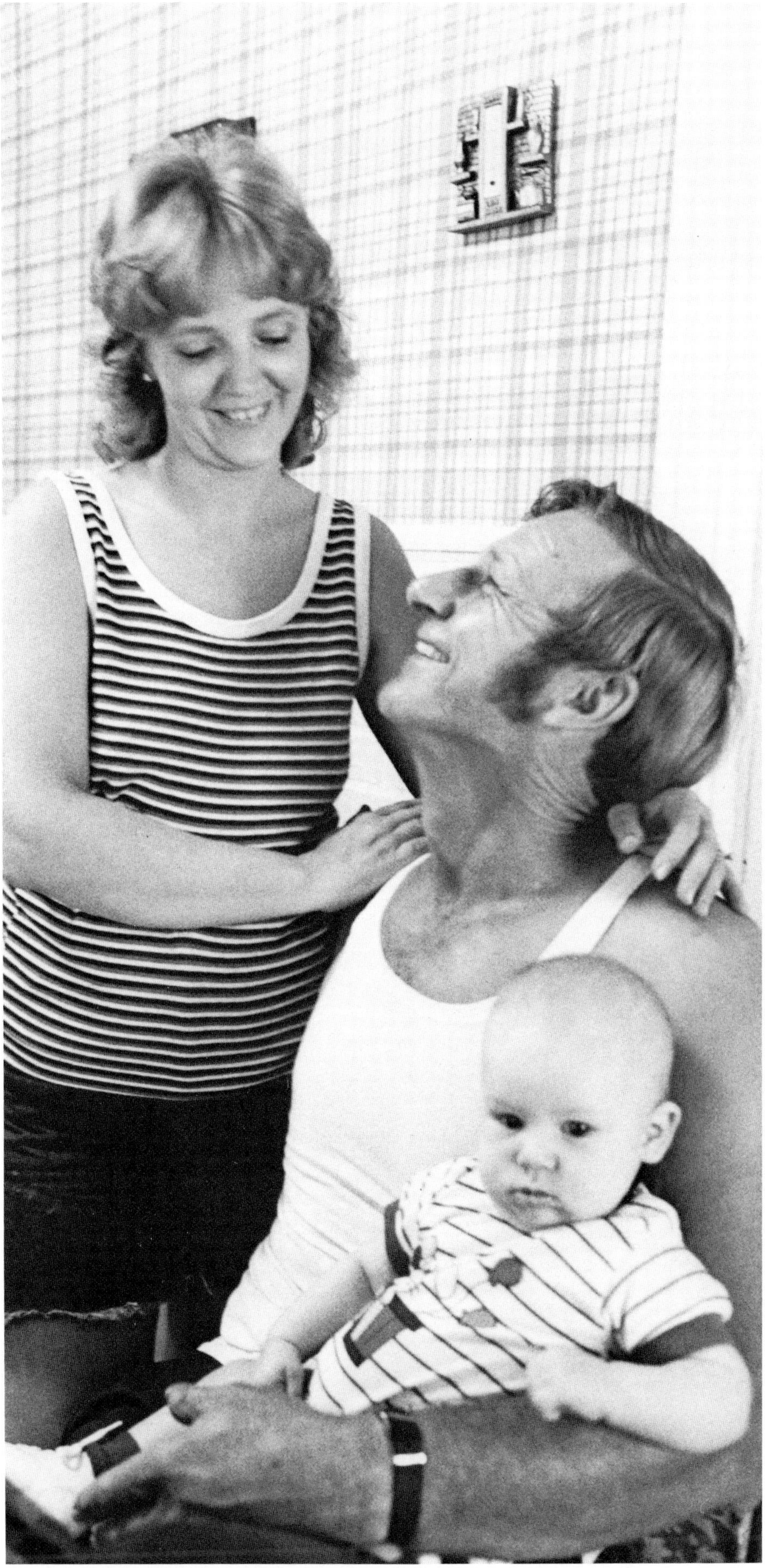

tended families, and the long history of coal mining struggles delineate the life style. "It's hard to get away from any tradition that we might have, regardless of what it is," says a Kentucky miner.

In the boomtowns of the West, the community accommodates a transient population. "There isn't a lot of togetherness. People are kind of independent out here," comments a miner in Craig, Colo.

The transient population is more likely to consist of construction workers, who come to these communities early in the boom to open the mines and build the power plants, only to move on to the next boomtown.

Coal miners are less likely to be transients. They come to these towns from other parts of their State or from other western States; a few easterners — estimated by one large mining company as 5 percent of the work force — also arrive.

For most of these mining families, the experience of pulling up stakes and moving to a new job in a new town also means starting all over with new friends, to be found among the many strangers.

Whether living in the East, Midwest, or West, miners can be characterized as family people — although the family size is shrinking. According to the 1922 U.S. Coal Commission, the average miner's family in West Virginia had 4.3 children; today's average is 1.6 children per family. The national average in 1975, according to the Bureau of the Census, was 1.5 children per family.

Time with the family is important to many miners. Top left: Miner John Basil, from Bridgeport, Ohio, with his wife Dawn and their child. (Wathen) Bottom left: Grant Town, W.Va., miner Gary Lee Fleming, with his wife Pearl Ann and sons Gary Lee, II, and Timothy Jermaine. (Wathen) Right: Miner Kenneth G. Howerter, his wife Linda, and their grandson Chad, from Lewiston, Ill. (Wathen)

Dealing with Job Insecurity

The history of coal mining is one of boom and bust, of strikes and layoffs, and life in coal country has evolved to deal with interruption of income.

Sue Cline, who lives with her coal-mining husband Jerry in Prosperity, W. Va., explains:

Jerry's from a mining family that sticks together. During hard times, Jerry's dad would always give us money to see us through. And even now, with a 2-month layoff predicted, they have already started doing little things like buying us extra food they happen to see on sale.

Another hedge against hard times is farming. Whether planting vegetables on terraced hills in Kentucky, raising hogs and chickens in a back lot in West Virginia, planting crops on large midwestern farms, or raising cattle on a western ranch, many miners work to insure that food will be on their tables even when there is not much money to buy it.

Farming is also a way to prepare for an active life after retirement.

A few miners even become entrepreneurs. Wilburn Bridgmon, for example, lives on a 40-acre farm in Walker County, Ala., where he has built a commercial chicken operation. He plans to retire after 38 years in the mines and to devote his full time to raising chickens.

In some mining communities in the Midwest, farming and

mining constitute the foundation of the economy and are often tied to one another.

In Montgomery County, Ill., for example, during the 1978 miners' strike, farmers gave sacks of groceries to the miners to help them get by. Many miners in the Midwest own or lease land for farming.

"The western coal miner is different from the eastern counterpart; he is less tied to mining," according to Jesse Mascarenas, a miner in Craig, Colo. If laid off, he claims, he will not necessarily stay in the area but will get another job, perhaps helping local ranchers with haying, cattle tending, or crop harvesting.

Many miners farm as a second occupation. Farming serves as a hedge against hard times, as well as a source of enjoyment, both during working and retirement years. Top left: Local union president, J. N. Braden, of Providence, Ky. (Wathen) Bottom Left: Miner Del Geiser on his farm near Oak Creek, Colo. (Corn) Top right: Willard France from Astoria, Ill. (Wathen) Bottom right: Miner Wilburn John Bridgmon of Jasper, Ala. (Murray)

Work Shifts

Louis Ketchum, a miner in Craig, works the following hours: Starting on a Tuesday, he works 7 days on a day shift followed by 2 days off, then 7 days during the 4:00 p.m. to midnight shift and 1 day off, then 7 days from midnight to 8:00 a.m.—called the "graveyard" shift in western surface mines and the "hoot owl" in the eastern underground mines. Mr. Ketchum enjoys 4 days off before returning to the day shift. Many mines follow this pattern of rotating shifts.

In Mr. Ketchum's case, the change in shifts is not too troubling because he and his wife have no children. Nonetheless, his wife had to take a less well-paying, part-time job because she wanted work that would accommodate his days off. More than 23 percent of the miners responding to the Living Conditions Survey say that their spouses work, and in those cases, or when there are children in the family, everyone must take into account the changing work hours.

A miner in Pennsylvania says:

We just more or less pass sometimes. If I work day shifts, I come home and she's gone. I'd be in bed when she comes from her work. Then I'd get up and go to work. When I work the afternoon shift, that's the only time we see one another.

Another miner explains:

Second shift [4:00 p.m. to midnight], *well, you can't come to work on half a meal. You have to have a big meal. And if all of your kids are in school, the wife has to make another big meal for them. On the second shift, you never see your kids. On hoot owl, I go home and sleep till 5. It's the worst shift.*

One week in three, these miners lead normal lives. "Let's face it, day shift is the ultimate for anybody. I call that 'human week,'" says one miner.

Ca. 1946

Coal Miners' Wives

Life in some company camps in Appalachia in 1946 was described in the Boone Report this way:

She has risen, before her husband, with the first clamorous notes of the alarm clock, to prepare his breakfast and thus commence a long, tedious round of household duties. . . . If there were but one device on the banner of the miner's wife, it would be the wash tub with crossed broom and mop. The dirt in the average mining camp makes it so.

Today, the life of a coal miner's wife is radically different. A neighbor describes Lenora Moss, who lives in Mallory Hollow, W. Va., in the heart of central Appalachia:

If she's not at home, you might try the school. Or, if she's not there, could be she's at the ballfield. I think she coaches her little girl's team. Oh, wait a minute, she's putting together a gospel singing program this weekend,

Left: During peak production periods many mines operate 24 hours a day—using three 8-hour shifts. Miners are placed on rotating shift assignments, which often disrupt family life. (Elliot)

Top right: Miner's wives, such as this woman from Walker County, Ala., had few work-saving conveniences during the 1940s. (Lee) Bottom right: Miner's wife Carolyn Hicks of Bloomfield, N. Mex., enjoys the same modern appliances used by other American housewives. (Murray)

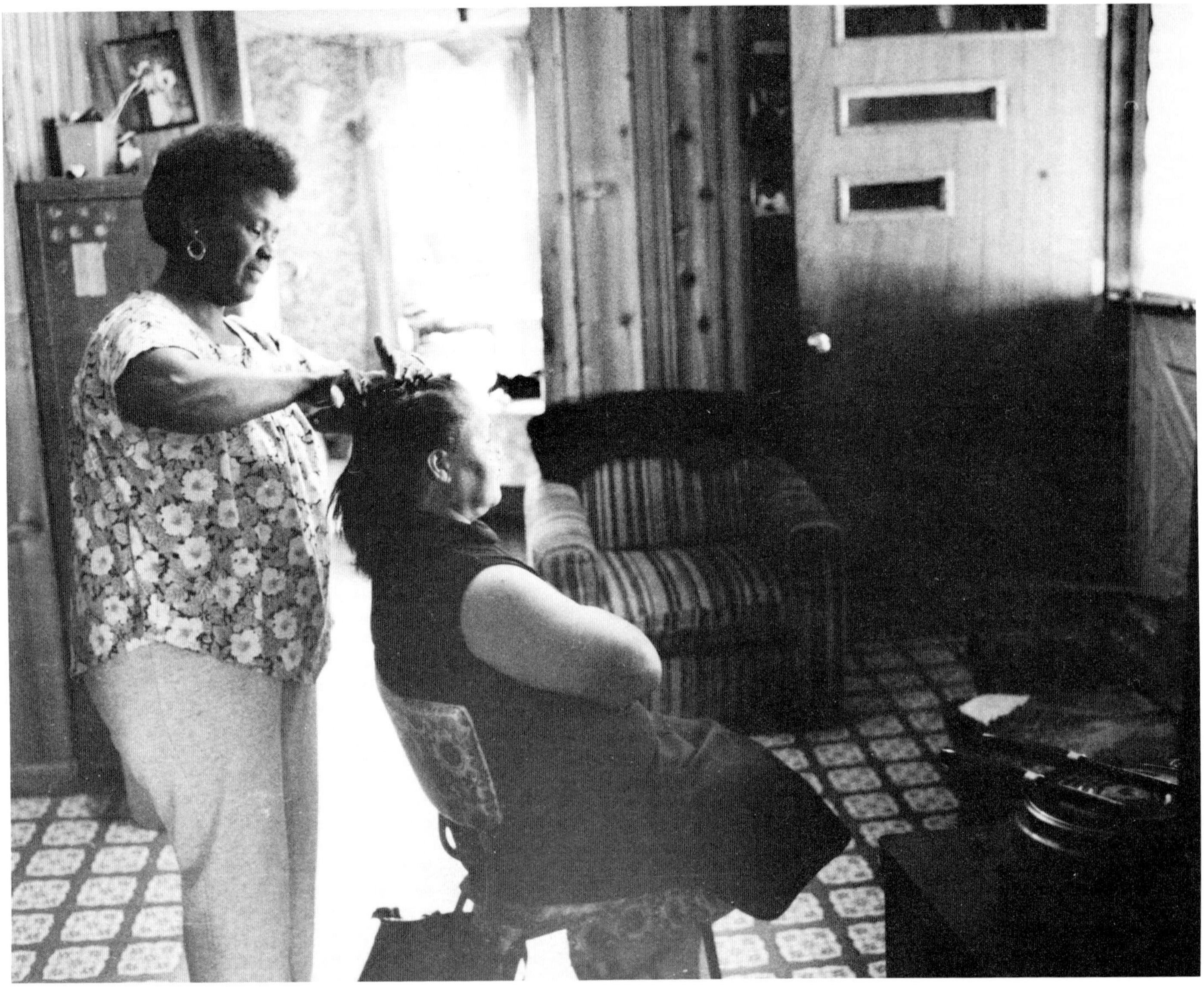

so you might try the church. Then again, her family's having a reunion from all over, so she might be at her sister's planning the food. What did she say she was doing today? It's not Saturday, so she's not fixing hair today. Well, if you can't find her, come back here and sit with me for a bit and she'll be home. There is one thing about Nora, you can always find her home when her kids get home.

In Appalachia, the life of a miner's wife revolves around her family and community, but it is no longer hopelessly lodged in the drudgery of daily housework. Like some other wives, Mrs. Moss works outside the home as a hairdresser, adding a second income to what her husband earns as a coal miner.

Mr. and Mrs. Moss have raised nine children, two of whom are still in elementary and high school. All the others have graduated from high school. Two have finished college and one is attending a State university. Their father completed 11th grade.

Not far away, in Oceana, W. Va., Debbie Tate, 24, wife of a coal miner, mother of a 2-year-old and pregnant, belongs to a newer generation. Debbie classifies herself as "a half-liberated woman," not satisfied with the choices available to her and wanting to work, to enjoy, and to get more education:

I'd hate to think I had reached all my goals by the time I was 24. I think it would be my responsibility with two kids to be with them. Being a parent is a very rewarding and enjoyable thing, but there is something in my mind that keeps telling me I'm only 24 and I'm going to be completely tied down because I'm going to have to quit my job and there is just nothing else to do around here.

Fewer than 5 percent of Mrs. Tate's high school graduating class continued to college. She studied accounting at Beckley College, a 2-year institution, and took a part-time job during her first year there. When the local bank offered full-time work, she dropped out. After 2 years at the bank, she still makes little more than minimum wage. "Women are not part of the business community," she concludes.

Seeking Opportunities in the Midwest

In the Midwest, the opportunities for employment and cultural activities sometimes exceed those in Appalachia. Younger women tend to work until they have children, and plan to return to work when those children grow up.

Shelley Wolter, mother of a 2-year-old, lives in Steeleville, Ill. She taught school for 2 years before she had her child and hopes to go back to teaching as soon as her daughter is of school age. She still teaches kindergarten 2 days a week in the Lutheran school in town and substitutes in the public school. Mrs. Wolter also seeks out other opportunities:

This area is culturally stagnant. There isn't much to do. We go to church on Sunday, get together with friends, and sometimes go fishing. Some of my friends and I got together and decided that if we want to go to dancing school, music school, go to the ballet or the museums, if that is important to us, we are not going to sit around and blame the lack on the town. We will get a carpool and go. We won't sit around and cry. Instead, we will do something about it.

Work opportunities for women in coal-mining communities are few, and when available they are usually low-paying. Left: Miner's wife Lenora Moss has filled her life raising nine children, participating in community activities, and working as a hairdresser in Mallory Hollow, W. Va. (Corn) Right: Miner's wife Debbie Tate, of Oceana, W. Va., would like to have more work, recreation, and educational opportunities available to her. (Corn)

Wives in Western Boomtowns
The coal boomtowns of the West pose different challenges for the miner's wife. Some miners' wives are newcomers, and they may find it difficult to gain acceptance in a community of long-time local residents, other westerners, and some migrants from the East. Many families live in large mobile home parks. Although friendships among wives develop, some women lack roots in the community and do not involve themselves in it, in part reacting to the resentment with which they were greeted.

In Hayden, Colo., estimated 1979 population 1,900, newcomers help mine the coal to fuel the big mine-mouth generating plant. One woman says:

Hayden wasn't friendly to us — we came right when it boomed. The people who had lived here a long time thought we were trash if we worked for the plant because they thought we wouldn't stay. The people who will be permanent are treated differently. Some people plunge right in and change the community. Some just stay drunk all the time and don't do anything.

There seldom is work for these wives except for a few jobs as waitresses in restaurants and bars and a small number of secretarial and clerical positions connected with mining. Besides, many wives do not need to work because their husbands earn as much as $30,000 a year. "I

hardly know any women that work. The women don't get paid as much as they should. If I worked, I would cause my husband to get into a higher income bracket and we'd pay more income tax," says a woman in Hayden.

Albert and Delores Henrie and their four sons live in a three-bedroom mobile home in Castle Dale, Utah. They moved in 1979 from Salt Lake City because, as Mr. Henrie explains, "I had to support my family and this was the only place I could make the money I need to do it." Neither Al nor Delores Henrie likes living in a mobile home. Mrs. Henrie says:

There's no backyard. No place where I can sit with any privacy. I'd like to feel a little bit of freedom. There's just no place for the kids. We've got four boys and there's just no place for them to play except on the road. When the owner of this trailer park was getting the permit for it, he said he'd put a playing field in and a playground. But he still hasn't done it.

Living in the coal boomtowns of the West often requires adjustments to cramped mobile home parks. Left: This mobile home park offers little room for privacy or space for children to play. (Murray) Right: Miner Albert Henrie, his wife Delores, and their four sons came to Castle Dale, Utah, in 1979, because of the high wages offered in the mines. They live in a mobile home. (Corn)

Earning and Spending

When the coal market is favorable and work is steady, the coal miner earns wages comparable to or better than those of other industrial workers. In many small communities, the miner is near the top of the economic strata.

A recent national news study comparing "gainers and losers in real pay" among 42 occupations, indicated that coal miners have the highest average weekly pay before taxes: $434.70. Farm operators, in that survey, earned an average $245.96 in weekly wages, school teachers $289.23, and clothing workers $149.46.

Coal miners have widened the gap between themselves and other industrial workers in the past two decades. Today, they rank with steel workers at the top of the industrial pay scale, ahead of workers in the automobile industry and substantially ahead of those in chemicals and all manufacturing; their earnings are about three times that of workers in textiles.

Since 1974, coal miners' wages have risen faster than those of the general labor force. Under the 1978 UMWA-BCOA contract, these wage rates will increase 30 percent during the 3-year life of the contract.

Based on 52 weeks of work including paid vacation, a miner earning $435 per week would earn $22,620 for the year. The estimated 1979 average income of the 1,281 miners responding to the Living Conditions Survey was $19,442.

The Survey indicates that miners in the West have the highest annual incomes—an estimated average of $24,055 in 1979. Forty-five percent of the western miners responding to the Survey earned at least $25,000 per year.

Miners in the Midwest, according to the Survey, have average incomes of $20,950. Average incomes in northern Appalachia are $17,786; in north-central Appalachia, $17,571; and in south-central Appalachia, $16,138.

The retired miner has less disposable income. Pensions from the UMWA Health and Retirement Funds provide $275 a month and comprehensive medical care under the 1950 plan. Under the 1974 plan, the average monthly pension is $480. A retired miner also receives social security benefit payments, which amounted to $292 per month for a retired worker with no dependents in August 1979. A retired miner with Black Lung re-

ceives additional benefits for that disabling disease; a Black Lung victim with one dependent receives $295 per month in Federal benefits.

Strikes and Layoffs

Although union miners' wages are set by contract, individual incomes vary because of overtime work, wildcat strikes, and layoffs.

Wildcat strikes, of which the coal industry has the highest incidence of any U.S. industry, can significantly affect income. For example, the Bituminous Coal Operator's Association reported that miners lost $139.5 million in payroll in the 11 months preceding the 111-day strike in 1977, and $9.9 million in the much more stable 11 months following.

Market conditions sometimes produce shutdowns. "You can't ever tell when there's going to be layoffs. If we get layoffs around here, folks are really going to be hurting," says a miner in Walker County, Ala., underscoring a concern prevalent in certain coal regions.

According to the Survey, the highest percentage of miners laid off in 1978 was in central Appalachia where almost 33 percent of those responding say they were off for some period during that year. The long strike of 1977–1978 undoubtedly contributed to this high figure. The lack of demand for metallurgical coal, used to produce coke for steel mills, caused many of the layoffs. Layoffs were lowest in the West, where 7 percent of the respondents report being unemployed during some period in 1978. Much western coal is "steam"

When the mines are open, miners earn wages comparable to workers at the top of the industrial pay scale; however, annual earnings are frequently affected by strikes, layoffs, and short work weeks. Top right: A Kentucky miner picks up his paycheck. Bottom right: Miners apply for unemployment in Beckley, W. Va. (Wathen)

Left: A retired miner in Helper, Utah. (Corn)

coal, used to generate electricity, and it is mined on long-term contracts that provide relative stability to the work force. In addition, many western miners are nonunion and continued to work during the 1977–1978 strike.

In addition, 52 percent of the miners in Central Appalachia claimed they had worked some short weeks last year. Again, the lowest figure — 8 percent — was in the West.

Some miners' spouses — about 27 percent of spouses of all married Survey respondents — also contribute to family income. In the Survey, the average annual income of spouses was similar in each region, ranging between $8,000 and $10,000. By comparison, 46 percent of all wives in the Nation work, earning an average annual income of $5,790, according to the Bureau of Labor Statistics.

Miners as Consumers

Miners in earlier times were willing to live in company housing and buy at company stores. Today's miner, according to J. Knox McConnell, president of the First National Bank of Keystone, W. Va., is a new breed. "They say 'I'm gonna have what I want,' and when they want something they get it. The miner today wants to start where his parents left off."

Like most Americans, coal miners shop locally for small amounts of food and other supplies needed on a daily basis, and travel to larger shopping centers for bulk or special-item purchases. Because many miners live in rural areas or on the fringes of larger communities, their travel times for shopping exceed those of suburban or urban dwellers.

Analysis of data from the Living Conditions Survey indicates that average travel times for miners from home to the local grocery range from 14 minutes in the Midwest to a high of 21

minutes in the West; areas of Appalachia fall between those extremes.

Regional differences show up sharply in travel time from home to large shopping centers, according to the Living Conditions Survey. Miners in northern Appalachia drive for 27 minutes, on average, while their counterparts in the Midwest travel for 36 minutes. The times in south-central Appalachia (51 minutes) and north-central Appalachia (60 minutes) suggest the distances that those people must drive. Western miners, attuned to driving long distances as a matter of course, report the longest travel times to large shopping centers — 74 minutes.

Shopping Facilities Proliferate

In the past few years, Beckley, W. Va., has seen these fast food chains open: three Burger Chefs, two Pizza Huts, and a Burger King, Long John Silver's, McDonald's, Ponderosa Steak House, Wendy's, and Western Sizzling Steak House. A K Mart also opened and one mall is under construction with another in the planning stage.

The old pattern of owner-operated stores located in small towns or larger commercial centers is changing. Big, nationwide companies are taking the techniques of mass merchandising and franchising to coal country.

The result often is more variety of goods for miners and

other local residents. The arrival of the big chains also may spell the end of smaller facilities, such as the "mom-and-pop" stores familiar in many rural towns. That displacement is particularly true in the East and in the Midwest.

In West Frankfort, Ill., a shopping center started just 5 years ago already is being expanded;

another mall is under construction 6 miles north of the city and two more are operating within 30 miles of town. Six fast food chains have opened outlets in the past 5 years.

"You name it, we have it," says the acting manager of the Chamber of Commerce in Gillette, Wyo., a major coal boomtown in the West. "This is the fast-food capital of the world." In 1978, with the population rising rapidly, a K Mart and a similar discount store opened for business; since then, six fast food chains have started operations and three more have announced their intention to locate in Gillette. Malls and shopping centers are opening in various parts of the city.

Top left: Twenty-seven percent of miners responding to the Living Conditions Survey say that their wives work. Far left: Miner's wife Roberta France of Astoria, Ill. (Wathen) As in other American communities, fast food chains, shopping malls, and supermarkets are now found throughout coal-mining areas. Top center: Downtown Smithers, W. Va. (Murray) Bottom center: Miner Carlton Meyers and his family in Jasper, Ala. (Murray) Top right: A traditional general store in Ky. (Wathen) Bottom right: A supermarket in Farmington, N.Mex. (Murray)

Borrowing

Because many miners earn substantial incomes, short-term credit is sometimes more available to them than to other workers. As one miner says of local banks:

You can tell them you're a coal miner and you can go down and buy a Cadillac. Tell them you want $10,000 to buy an automobile, you get it right now.

Coal miners make frequent use of installment buying, claims Ronald Peterman, Manager-Director of the UMWA Federal Credit Union. The credit union is not financially affiliated with the UMWA; membership is open only to current members of the UMWA and their families and to widows of UMWA members.

The credit union plans to open 24 new branches within the next 3 to 5 years, and expects its current membership of 26,000 to expand to 100,000 within that period.

Ca. 1946

According to Mr. Peterman, the credit union made 25,000 loans during the 111-day strike in 1977–1978, mostly for food and emergency needs. Ninety-eight percent of those loans, he says, were paid within a year.

Company Stores

The company store, a coal country institution from early in this century until the late 1940s, was the miner's main source of food and supplies.

Coal companies owned and managed these stores, usually in the company town or near the mine site. Miners and their families commonly shopped in the company store either because they did not have the transportation to travel or because the store extended credit that they could not obtain elsewhere.

In 1947, there were 2,850 coal company stores in the Nation. Today, only a handful survive and, because miners have become mobile and consumer-conscious, even these must compete with other stores and shopping centers.

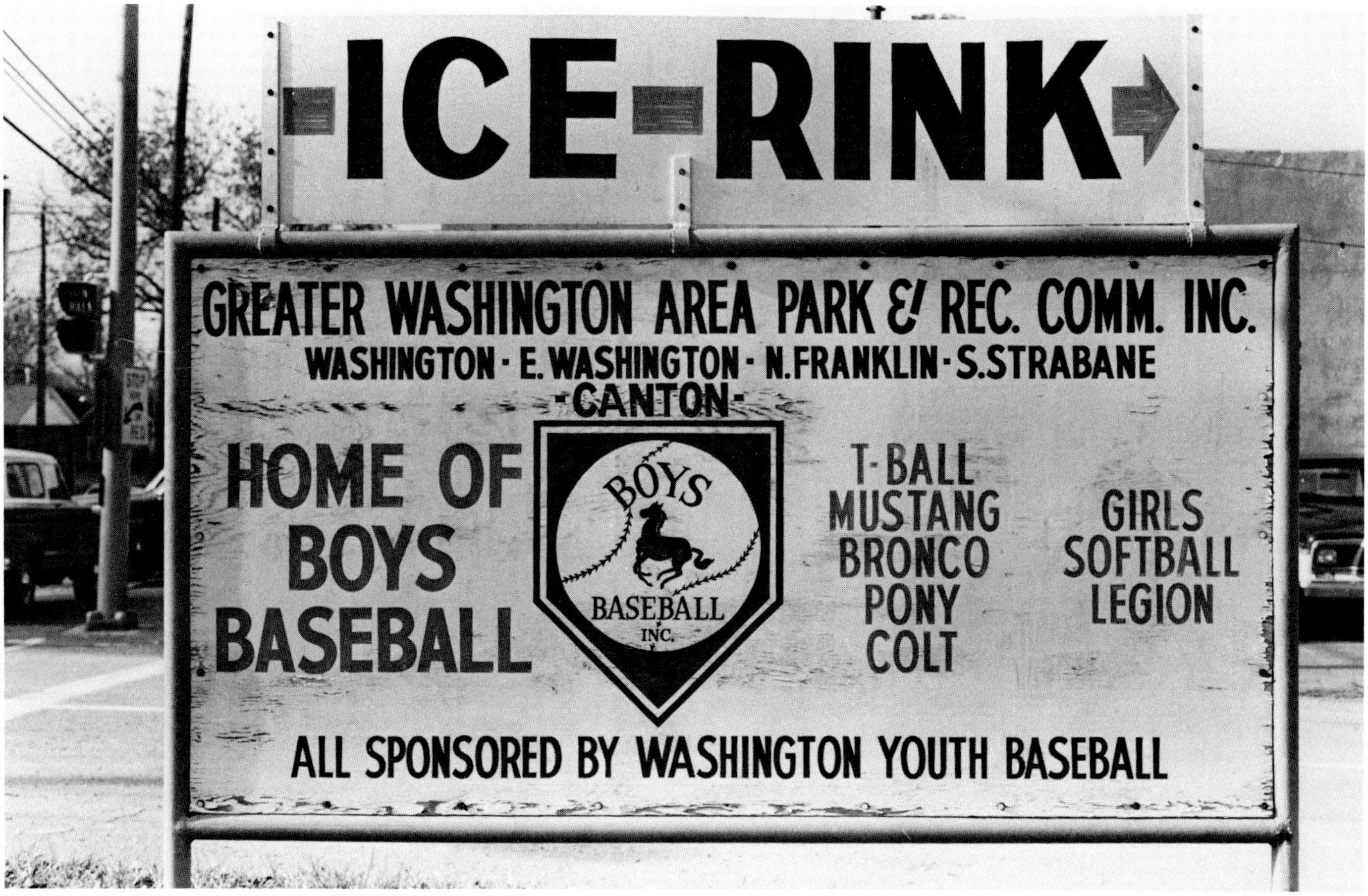

Enjoying Leisure Time

Coal miners work hard, and most of them look to leisure time and vacation as opportunities to forget the job, enjoy family and friends, and engage in various recreational activities.

Availability of facilities varies widely. In some parts of Appalachia, for example, mining families complained of lack of facilities. In other areas, such as Washington, Pa., however, the situation differs. One miner's wife explains:

There are nice parks here with three public schools. Washington Park has a pool, a new ice-skating rink, tennis courts, a children's playground, a baseball field.

Her husband adds:

This isn't the most glamorous place in the world, but it's a good town to raise your family in. There are movies, nice clubs, good restaurants. In the summertime you can go to the park. Everything we need, we have right here. I go water-skiing, boating, fishing along the river. People say that it's a small town, but there's lots to do here. We have a YMCA and a YWCA and they have swimming and sports for all the kids.

Coal miners' preference for such outdoor activities as hunting, fishing, and camping appears to be common in all regions. Unlike man-made and managed recreational facilities, the woods and water can accommodate the miners' varying work schedule. A miner from Kentucky reports that: "Almost everybody in these mountains owns a gun, or five or six." A miner from Washington, Pa., adds: "The first day of deer season, you might as well shut the mines down. Everyone takes a day off and goes hunting."

Fishing and boating are popular with miners. Al Giles, whose butcher shop sits on the banks of the Monongahela River, bordering Washington County, Pa., observes:

In the summer this river is full of miners and their ski boats. That marina down the road has about 150 boats and half of them belong to the miners. These miners like to spend their money on their speed boats and on fishing equipment.

Thirty years ago, the only credit miners could obtain came from the company store. Top left: Today, miners obtain credit from places such as this bank in Canton, Ill. to purchase cars, household appliances, and furniture (Wathen) Bottom left: Company stores, such as this seen in 1946, have become all but extinct. (Lee)

Right: Although all coal-mining communities are not as fortunate, Washington, Pa., has a full complement of indoor and outdoor recreational facilities, including tennis courts, swimming pools, and playing fields. (Owre)

72
ZERO

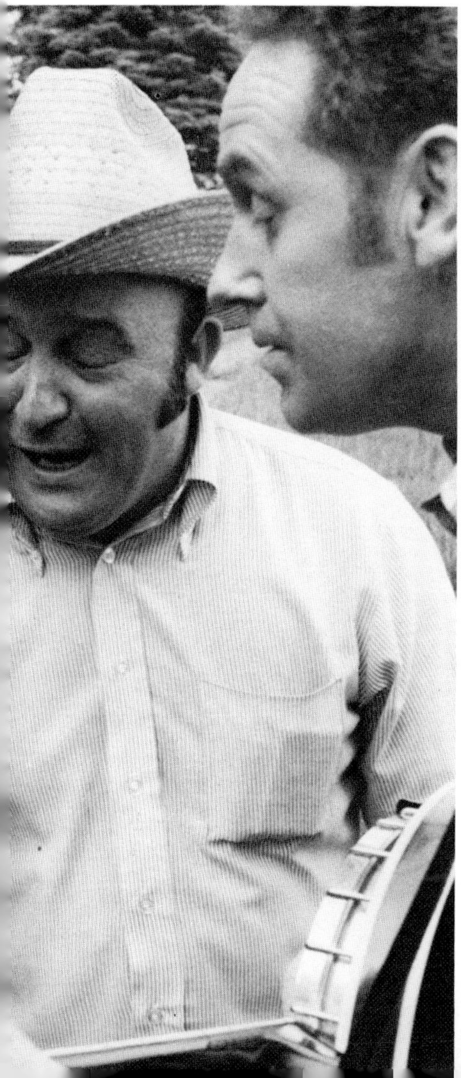

The availability and variety of recreational activities varies from one mining community to the other. Some communities have little more than bars and pool halls, but others have much more. Top left: Mountain Motor Speedway, in Isom, Ky., is owned by a mine operator. (Wathen) Top middle: Roadside in Harlan, Ky. (Corn) Top right: Elmer's Tavern in Millersburg, Ind. (Simmons) Bottom left: Disco in Raleigh County, W. Va. (Wathen) Bottom middle: A bluegrass festival in Coalwood, W. Va. (Corn) Bottom right: The Sons of Italy hall in Cokeburg, Pa. (Wathen)

BATHING
ONLY WHEN
LIFEGUARD
ON DUTY

Vacations

Many miners go camping and some own campers equipped with television, ovens and bathrooms. A Pennsylvania miner notes:

I have been spending the last few vacations in my camper in the State park. Usually I go to just one campground, as opposed to traveling to two or three places. The kids get to swim, ride bicycles, sit around the campfire at night—just more or less relaxing.

About 76 percent of the miners responding to the Living Conditions Survey say they travel away from home during vacations. In the West, the percentage was highest, with 86 percent saying they leave home.

The coal market often determines the range and nature of the miner's vacation. A miner in St. Clairsville, Ohio, explains: "Whether you go on a vacation or not depends on how you are working. If you work steady the month before, maybe you go away. If you don't, you stay home. You can't plan a vacation. You don't know if you are going to have any money for a vacation."

Vacations, some miners claim, are a necessity.

There isn't an occupation today that needs more time away from the job, except maybe an air traffic controller, than a coal miner. We're in dust, it's damp, it's cold, it's dark. We need more time off. Dust is dry, it gets on your skin, pulls the oil out of your hair. All of us have rough skins, we need a break from that.

Perhaps because of the long hours spent in the dark, below ground, miners seem to have a special fondness of outdoor activities. Top left: Deer hunting near Harlan, Ky. (Corn) Bottom left: A beach, built on the site of an old surface mine in Warrick, Ind. (Simmons) Top right: Miner John Alex Taylor III and his son Jeff, from Rupert W. Va. (Wathen) Middle right: Camping near Beulah, N. Dak. (Murray) Bottom right: Rupert, W. Va. (Wathen)

Education in the Coalfields

The coal miner today bears little resemblance to the illiterate immigrants of the 1920s who were recruited from central and southern Europe to mine the anthracite of Pennsylvania or the bituminous coal of Kentucky and West Virginia.

The miner operates complex machinery, works within a network of safety laws and regulations and follows complicated labor negotiations. As parents, miners desire sound education for their children.

Miners are better educated than in the past, and the educational level appears to be rising among young miners. Miners also are better trained than ever before, although classroom training, in conjunction with on-the-job training, is comparatively new and not yet universal.

Educational Levels of Miners

A public school teacher in Big Stone Gap, Va., describes what is required of young people in the coalfields who are contemplating mining as a career:

You just have to have a high school diploma to work in the mines. The whole thing is mechanized, and you have to use your brain or you'll wind up getting squashed on the floor someplace. And they have so many people with high school diplomas and college degrees that they can pick and choose. That's what we try to get across.

The teacher's statement is borne out by statistics. According to a UMWA survey in 1975, more than 60 percent of young people (24 years old or younger) entering coal mining are high school graduates, and almost 22 percent have some postsecondary education.

The median educational level of all coal miners has shown a steady, if undramatic, rise. According to one Federal labor force sample, the median years of schooling completed by coal miners was 7.5 in 1940, 7.7 in

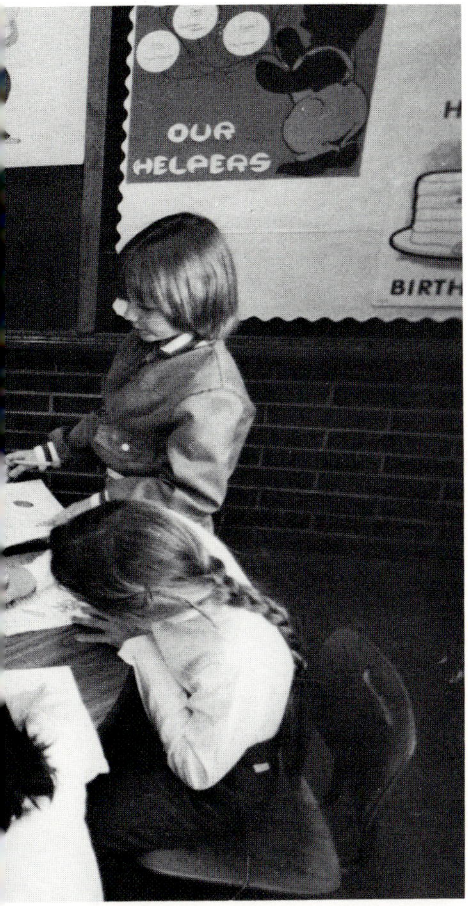

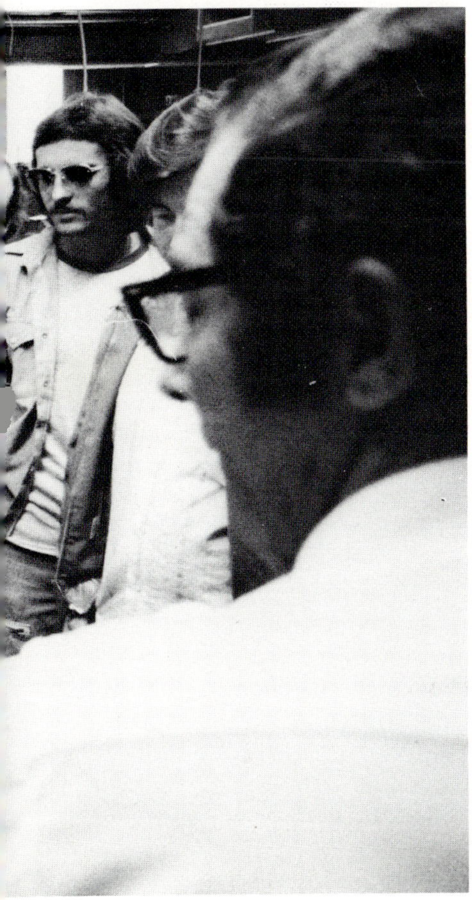

1950, 8.0 in 1960, and 8.1 in 1970.

Educational levels are higher among younger miners and among miners living in northern Appalachia and the Midwest, according to responses to the Living Conditions Survey.

Survey results show that 60 percent of miners 25 years old or younger have a high school education; of older miners, 59 percent of those 26 to 35, 49 percent of those 36 to 49 years old, and only 34 percent of those more than 49 years old have finished high school. While 10 percent of miners 25 or younger have at least 2 years of college, only 3 percent of miners 49 or older have reached that level.

The average level of scholastic achievement of coal miners remains somewhat below that of the population as a whole. The average for all respondents to the Survey was 11.4 years of school; the average for all adults 25 years old or older, according to the National Center for Education Statistics, was approximately 12.5 years in 1978.

Training a Miner

At the time of the Boone Report in 1947, all training was based on a buddy system. An older and experienced miner—often the father or other relative—trained an apprentice miner on the job.

Vestiges of this system remain today. A newly hired miner undergoes health and safety training and some task training before entering the mine. But the basic instruction still occurs on the job.

Oldtimers say that experience is the best teacher in mining; that only many years underground can train the ear to hear a roof that is "working" or about to fall.

Experience doubtless is invaluable in underground or surface operations, but current Federal law adds an element of classroom training in health and safety. These requirements were contained in the Federal Coal Mine Health and Safety Act of 1969, which later was significantly amended to become the Federal Mines Safety and Health Act of 1977. This law mandates that coal operators train inexperienced miners for 40 hours before sending them underground or 24 hours before starting them in surface work. All miners must undergo 8 hours of refresher safety training per year and task training before reassignment.

In addition, the National Bituminous Coal Wage Agreement of 1978 requires that inexperienced miners be prohibited from operating heavy equipment near the face (the exposed coal seam) for 45 days after completing the mandatory health and safety training program.

State laws are important in establishing minimum requirements for skill attainment and training. One of the most progressive laws is the West Virginia Miner Training, Education, and Certification Act. This law mandates 80 hours of classroom training for prospective underground miners (40 hours for surface miners), and requires successful completion of written examinations to demonstrate an understanding of safe work procedures.

The first signs of a more formal approach to educating the miner as miner seem to be appearing in secondary and postsecondary school systems in the coalfield States.

Education in the coalfields has improved considerably since the time of the Boone Report in 1946. Top: A school in Craig, Colo., a coal boomtown. Bottom: The Willis School in Logan, W. Va., is one of an increasing number of secondary schools that offer courses in coal mining. (Corn)

Learning Mining Through the School System

For many years, the secondary and vocational schools of the United States have offered training in such fields as automotive mechanics, basic carpentry, and home economics, but until about 7 years ago, few schools offered basic coal mining to students.

Today, a number of school systems offer the training required for a young man or woman to enter the occupation. Students completing these courses can enter mining.

After high school, the aspiring miner can begin work in a mine immediately, earning standard scale, which is $63.39 per day for a trainee.

In an increasing number of regions, the working miner—even an entry-level miner—can continue education in mining by entering a postsecondary school. These often are vocational or technical schools or 2-year community colleges that offer courses in advanced coal mining.

Students most likely to enter these schools are those interested in becoming foremen or other supervisors. A miner must have 5 years of experience and, in most States, pass an examination in order to become a foreman. Years ago, only State mining departments offered training for someone wishing to become a foreman. Many still do, but those courses traditionally are narrower in scope than the courses offered by vocational and community colleges.

The postsecondary training in mining is designed to accommodate the miner's schedule. Many institutions allow students to work 6 months and study 6 months; others offer schedules that accommodate the student's shift at the mine. A number of institutions give courses on company property at the mine, and a growing number of companies now pay a portion of the tuition. Credits at many of these institutions are transferable to 4-year college programs.

Currently, 30 postsecondary schools across the country offer courses in mining technology. Three are in Arizona, one in Colorado, five in Illinois, two in Indiana, six in Kentucky, one in Ohio, two in Pennsylvania, one in Utah, two in Virginia, five in West Virginia, and two in Wyoming.

One of these is Pikeville College, in the heart of eastern Kentucky coal country.

The college was founded in 1889 as a Presbyterian institution devoted to training teachers and emphasizing the liberal arts. Today, having dropped its religious affiliation, the college has added a 2-year degree program in mining technology, and is training more people for mining careers and fewer for teaching.

Pikeville College has 500 full-time students and 200 part-time students for the 1979–1980 academic year, and 125 are enrolled in the 2-year mining program.

Since 1973, 400 students have

graduated from the coal mining program and 91 percent hold jobs in the coal industry — primarily management.

The first year curriculum includes courses in Coal Science (geology), Techniques of Supervision, Evolution of Health and Safety Regulations, Mining Mathematics, Safety and First Aid, Mine Maps, and Mine Rescue. During the second year, students take courses in Mine Roof Control, Electricity, Ventilation, Preparation Plant Technology, Blasting, Mining Math and Mining Law (which includes study of labor contracts), Labor Relations, and the Coal Mine Health and Safety Act.

Students can join a summer internship program that allows them to hold a job at a coal mine. A budget of $280,000 sustains the program, 20 percent of which comes from the college and the rest from the U.S. Mine Safety and Health Administration and the Kentucky Department of Minerals and Mines.

Those enrolled — including 15 women — are mostly from coal-mining families in eastern Kentucky and southern West Virginia; many of the students work in the mines during the day and take courses at night.

A western counterpart of Pikeville College is Casper College in Casper, Wyo., which is a 2-year institution founded in 1945. Casper College currently offers courses in coal mining, with an emphasis on surface operations.

Experienced miners can continue their education under certain new programs, even as prospective miners study the basics of the industry. Left: Ray Bob Goodman, of Lorado, W. Va., 37 years a miner, goes to foreman's class in surface mining after working the "hoot owl" shift, which ends at 8:00 a.m. Top right: Pikeville College, in Pikeville, Ky., offers a 2-year program in mining technology, and allows students to work in the mines while studying. Bottom right: One of 15 women currently enrolled in the program at Pikeville College. (Corn)

Advanced Study in Coal Mining
Degree courses in coal mining are offered at 22 colleges and universities in the United States, including such institutions as the Colorado School of Mines, the University of Kentucky, the University of Missouri (Rolla Campus), Pennsylvania State University, the University of Pittsburgh, West Virginia University, and Virginia Tech Institute.

The Colorado School of Mines was founded in 1874, before Colorado became a State. Today the school has an enrollment of 2,845 students, including 461 women. (The first woman graduated from the school in 1898, but by 1969, only nine had completed the undergraduate course.) Students are drawn from throughout the United States and about 50 foreign nations.

The school offers courses in both coal and hardrock mining (as well as petroleum engineering and other subjects), on the undergraduate level, and admits qualified candidates for the master's degree and doctorate in mining engineering.

Graduates interested in coal mining often gain some experience in production, either during summers while students or after graduation. This gives them practical knowledge of underground or surface mining. Some remain in production, aspiring to become foremen or shift bosses. Others go into mining engineering with the coal company or, on occasion, the parent company.

Educating Miners' Children
Lack of adequate financing, poorly trained teachers, emphasis on athletics at the expense of learning — this has been the scenario for education in the Appalachian coalfields, particularly Kentucky, for decades.

Poorly educated, untrained in skills that would allow work options, the coal miner characteristically was a person trapped in a life cycle he could hardly alter. A 64-year-old retired miner in eastern Kentucky reminisces:

I looked out there and this is the way I looked at it. There is nothing. No future. Nothing. Will it always be like this? I thought.

Now conditions are more attractive. Now you have machines instead of you being the machine. The young fellow comes out of high school. He wants to get a job in the coal mines for a while because he wants to work enough to go to college. He gets in there, he works a bit in the mines, he gets some money and he goes to college. But where does he go? Back to coal mining because now he's got a college degree and thinks, "There's advancement for me up there."

Educational opportunities for youngsters in the coalfields clearly have improved since the time of the Boone Report. But the degree of improvement ap-

pears to vary considerably, depending on the State, county, and local school district.

This report makes no attempt to analyze the quality of public education for miners' children that exists in coal communities in 1979. But some characteristics and historical changes were observed during the course of the study, and these are reported here.

Public schools in coal country often are consolidated — a process begun more than 30 years ago — and offer at least the basic courses and services to children. This situation represents a major change from the days of the Boone Report, when one-room schoolhouses in coal company towns were the rule.

Comments by miners and their families make it clear that public education in the coalfields suffers many of the same problems experienced elsewhere in the Nation. Parents complain about schools being too big and impersonal, too costly, too lax on disci-

Left: Modern coal mining is a subject of interest to textbook publishers as well as to institutions of higher learning; 22 colleges and universities now offer courses in this subject. (Corn)

Right: At the time of the Boone Report, many miners' children attended one-room schools, unlike this 1979 classroom is in Stoco Elementary School in Raleigh County, W. Va. (Wathen)

pline, too anxious to please students, and not vigorously teaching the three Rs.

A popular — although easily misleading — measure of the soundness of education involves comparative scores on the various tests used for college entrance. The most common tests are the Scholastic Aptitude Test (SAT) and the American College Testing (ACT) test. Scores from these tests do not constitute any definitive measure of the quality of a school system; they merely indicate what some students from that school system score, and how those scores compare with the national average.

Moreover, different institutions of higher learning require different tests. Students aspiring to enter State institutions in their own State, for example, might find that they must take the ACT test, which is preferred by many public institutions. Other students might take the SAT for entrance to a college or university requiring that test.

Average scores for a sample of five coal counties indicate that those students who take the test score in the middle ranges. (See Table 1.)

Table 1.
Average 1979 Scores on the Scholastic Aptitude Test (SAT) and the American College Testing (ACT) Test

Reporting Unit	SAT Verbal	SAT Mathematics	ACT
National	429	468	18.9
Moffat County, Colo.	510	610	20.0
Franklin County, Ill.*	474	529	19.0
Campbell County, Wyo.	472	501	--
Pike County, Ky.**	470	450	15.5
Raleigh County, W. Va.	433	480	17.7

**School officials report that few students take the SAT test.*
***The SAT test was administered in Pike County for the first time in 1979; about 30 students took the test, according to school officials.*

Sources: National data are from the Educational Testing Service. County data are from county school officials.

Education in the West

Schools in coal boomtowns of the West often suffer the strains of rapid expansion. Students in some boomtowns attend newly built and well-equipped schools where teachers' salaries and per-pupil expenditures are among the highest in the country. Other towns find themselves with overcrowded schools, small budgets, and underpaid and demoralized teachers.

Some communities have been unable to build new schools because voters hesitated to approve bonds to build schools for a population increase of uncertain duration.

Carbon County, Utah, for example, is reluctant to proceed

with a school building program because of the uncertainty of the student population. The school system relies on portable classrooms.

The county also finds it difficult to plan for future growth because it is not yet sure if a nearby mine will open and when. If the mine opens, the town would see an increase of between 600 and 1,200 families, which would require a new high school.

The quality of public education in the coalfields varies by State, county, and local school district. Many school systems use contemporary equipment and techniques. Left: Science class at Independence High School in Raleigh County, W. Va. (Wathen) Top right: Wagonwheel School in Gillette, Wyo., a coal boomtown that planned for the influx of miners. (Berger) Center right: A teacher at Stoco Elementary School in Raleigh County, W. Va., holds class outdoors. (Wathen) Bottom right: The woodwind section practices at Oak Creek, Colo., High School. (Corn)

In Emery County, Utah, one high school is so overcrowded that partitions have been built in the cafeteria to create three temporary classrooms. Two classes meet in the library, one on the auditorium stage, and another in a chair closet.

Gillette, Wyoming

Gillette has planned ahead.

Gillette's experience differs from that of other boomtowns because its growth has not been unanticipated. Besides, it offers some of the highest school teachers' salaries in the country; a beginning teacher earns $12,700, and the maximum on the pay scale is $24,712.

The entire school population has increased from 3,400 in 1975 to 4,400 in 1979, and school budgets have risen proportionately.

The town has one high school, one junior high, and eight elementary schools, with two more under construction. The high school was built in 1971 for a capacity of 1,400 pupils; in 1979, it has an enrollment of 928. The junior high school was rebuilt in 1974 for 1,100 students and has an enrollment of 1,000. The eight elementary schools have a total capacity of 3,000 and enrollment of 2,487.

These numbers foretell that the elementary school population will soon surge into the high school and junior high school. When it does, the town will be ready because, according to Raylene Suchor, Assistant Superintendent of the Campbell County School District, financing is available for new construction. The district has authorized $32 million for the 1979–1980 school year with $12 million designated for construction. A new junior high school is planned and will be financed by a recently passed bond issue.

The main administrative problem in the Gillette schools is the widely fluctuating school population. For example, in a single week in 1978, 36 new students registered and 33 left.

Campbell County, where Gillette is located, is a sprawling area of 5,000 square miles and 21,000 residents. The total county budget is $13,681,068, of which none goes to the schools and hospital. Schools are supported by the coal severance tax. Last year $4.2 million was collected in severance taxes of which 87 percent went to the schools. There are now 4,903 students from kindergarten through grade 12 in the county. Per-pupil expenditures are $2,500.

Education in Appalachia

In contrast to Campbell County, with its large area, low density of population, and considerable revenues, stand counties such as Letcher County, Ky., which is smaller, more densely populated, and relatively worse off in terms of revenues from coal severance taxes or any other source.

Letcher County, located in the southeastern corner of the State, covers 339 square miles and has a population of 28,000. The two school systems within the county, with 6,441 students, are supported by State and local funds and have a budget of $6.5 million, which comes to $1,000 per pupil per year. A small portion of the severance tax is returned to the county, none of it designated for the schools.

Teachers' salaries run below those in Gillette; in the Jenkins Independent School District, located in Letcher County, the minimum teacher salary is $9,660 and the maximum is $16,990.

What Some Miners Say About Appalachian Education

Some miners say that their children are being educationally deprived. A miner in southwestern Virginia says:

My main concern right now is my kids. I'd like to see things so much better for them, and the school system is pathetic. There's nothing really that I could do about it except leave, and I hate to leave. I'd sure like to see it changed.

Although the quality of education varies from community to community in central Appalachia, and a number of new schools have been built within the past few years, miners and their families express anger about schools they deem substandard.

"Did you see those little red plywood buildings as you drove through . . . ?" asks one mother. "You wouldn't believe it to go down and walk through that school. In the winter time those buildings have to be cold. I know, because I went to eighth grade there."

A teacher in an eastern Kentucky school district remembers her impression after visiting an elementary school in a nearby town in 1977:

When I walked in, there was a potbellied coal stove with a fence around it so the children wouldn't get burned. At rest time, little children got down on their mats and when they got up their little hands and little faces were just as black as they could be with coal dust. I think it was the only school in the county that had those conditions. But the parents picketed there less than a year ago and they've decided to build a new school now. Somewhere they came up with the money. They said all along they didn't have the money.

Conditions are not uniform throughout the State. School facilities in western Kentucky are noted by one parent as "real neat. Why don't they have schools in eastern Kentucky that are decent and the children would be proud to go to?"

Education in the Midwest

"In my opinion, we have one of the finest educational systems in this area."

"Where my kids attend school, the teachers seem like they are very dedicated."

"We are really well satisfied with our schools. We have good teachers, new buildings. We're better off than the big cities. We've got everything they've got except their problems."

These comments reflect the views of miners, foremen, and community people in Benton, Ill. One miner's wife noted, "The schools are good. I guess you heard about the national testing they did for the three Rs. Well, Pinckneyville was running 1 to 2 years ahead of the national average."

Nearby towns of Sesser and Vale have recently built a new consolidated school that accommodates 1,100 students, from kindergarten through high school. Thirty-five percent of all graduates go on to junior college or college. Jasper County, Ill., some local miners contend, has the best schools and educational opportunities in the region.

Some parents in these regions are actively involved in the schools. Sandy Kershaw of Hillsboro, Ill., the wife of a miner, worked hard to get a bond issue passed in 1978 so that a new school could be built. "We went door to door trying to get people to vote for it. I even wrote an article for the paper on it," she says.

Linda Hinds, a coal company employee, lives with her husband Ed and 18-year-old son in the town of Witt, Ill. Mrs. Hinds has been a member of the school board for 4 years and was on the city council.

Jim LaVault, of Sesser, is the father of three children and has worked as a repairman in a mine for 9 years. He is also finishing his second term as a school board member.

At the same time there are several counties where schools have problems keeping up educational standards. The cause often is lack of funds to build new schools, but even then the picture is far from bleak. Some miners say that, despite poor facilities, teacher morale is high, and the children are well taught. In addition, several of these midwestern towns have vocational high schools and nearby community colleges.

Coal mining is never far from the homes or schools in the coalfields. In this unusual situation, the Big Rock School in Big Rock, Va., is located in the middle of a coal mine. (Murray)

Lifestyles of American Coal miners have changed dramatically over the years since the Boone Report. Some recreational pursuits of modern miners were probably undreamed of by their miner forebears. Yet traditional values remain strong forces. Top left: Miner George Austin of Greenville, Ky., is a student pilot. (Wathen) Bottom left: Tony Meade, his wife Sheri, and their children, of Hunt, W. Va. (Corn) Top center: Miner Douglas Joe Harris of Madisonville, Ky. (Wathen) Bottom center: Members of the North American Coal Company golf league at Powhatan Point; Ohio. (Wathen) Top right: The 26th annual reunion of the coal-mining Braden family of Providence, Ky. (Wathen) Center right: The Sophia (W. Va.) Little League team, composed mostly of miners' children, whose coaches and officials are miners. (Wathen)

Women In the Coal Mines

She had worked in factories, driven school buses, and found part-time work in stores, unable all along to earn enough to support herself and her four children. She applied for work in coal mines because wages were higher, and "got the cold shoulder at every company except Beth-Elkhorn." She persisted. "I went back and literally begged them for the job."

Her perseverance paid off, and Sandra Bailey was offered a job in 1975 in a Kentucky mine. On September 12, dressed in layers of baggy clothes, she went into the mine, was handed a shovel and told, "Here, shovel, shovel 'til we come after you." The labor and the command were prosaic, but to that woman and others who followed, being allowed to shovel in a coal mine was a victory.

A Revolution in 1973

Five years ago, a revolution occurred in the coal-mining work force when women began to take their place alongside men in mines across the country.

Prior to 1973, government records show no women miners. A few women may have mined coal as occasional workers in small family operations or for brief periods of time during and shortly after World War II. The National Institute of Safety and Health reports that no women took the required miners' examination prior to July 27, 1973. In 1977 there were 992 women and in September 1978 the ranks had grown to 2,000. In 1979, there are approximately 2,600, constituting 1 percent of employed miners.

Overleaf: Until 1973, few women worked in coal mines. Now 2,600 underground miners are women. One is Marge Mundorf, mother of a 3-year-old, who mines coal in Clyde, Pa. (Simmons)

After being turned down several times, Sandra Bailey of Pike County, Ky., was hired in 1975 by the mine where she now works as a repairman first class. Bottom: Sandra Bailey with her children. (Corn)

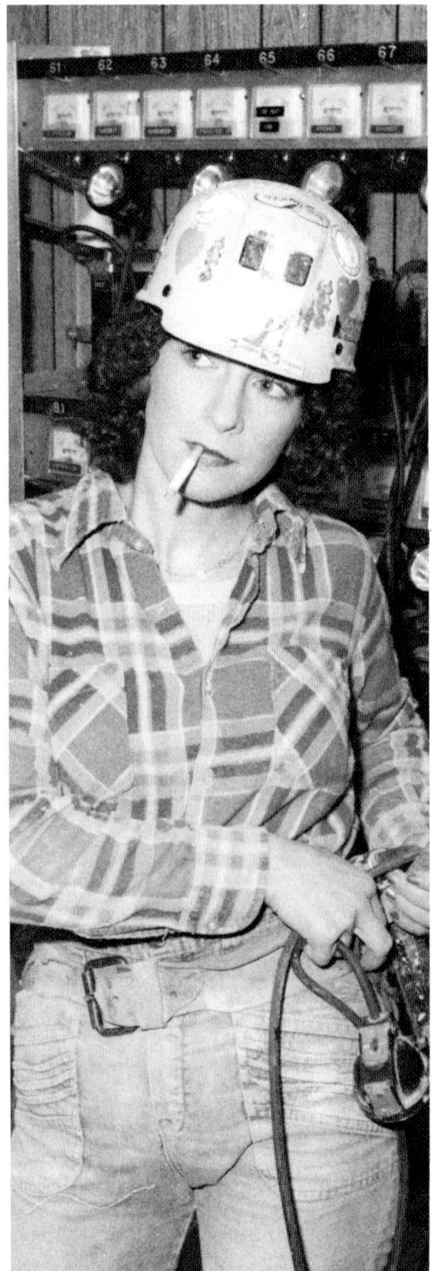
61 62 63 64 65 66 67

An early stimulus to integration was a 1965 Executive order issued by President Lyndon B. Johnson, which barred discrimination in employment on the basis of sex and race and required businesses receiving Federal contracts to develop affirmative action plans for minorities. The Executive order was based on the Civil Rights Act of 1964.

Grassroots campaigns against sex discrimination in the coal industry eventually followed. One important catalyst has been the Coal Employment Project (CEP). This Tennessee-based group received an 8-month grant from the Women's Bureau of the U.S. Department of Labor in August 1978 to develop a model training program for women who want to begin coal-mining careers.

In August 1979, 20 women completed the training program and began the process of being placed. Because CEP received more than 100 applicants for the original 20 slots, it was considering continuing the pilot project.

The group also has filed lawsuits against coal companies that it believes have violated Federal prohibitions against sex discrimination. In May 1978, the organization charged 153 coal companies with sexual bias and in December 1978, one major coal company, Consolidation Coal Company of Pittsburgh (the second largest in the country) agreed to pay $370,000 to 70 women who were denied jobs. It also agreed that one out of every four miners it hires will be a female.

After struggling to enter coal mining, women now represent 1 percent of the work force. Among them, clockwise from top left, are: Rose Mary Gamble, Jasper, Ala., (Murray); Doris Allen and Maybelle Gibson, Jasper, Ala., (Murray); Sheila Jackson and Patrisha Pugh, Jasper, Ala., (Murray); Zella George, Chandlersville, Ohio (Wathen); Helen Britt, Harlan, Ky., (Corn); Annette Plawer, Vermont, Ill., (Wathen); Joan Thompson, Hickory, Pa., (Wathen); and Kay Bennett, Peoria, Ill. (Wathen).

Why Women Enter Mining

Soft spoken, slightly built, mother of two, Linda King works in a deep mine outside Big Stone Gap, Va. Mrs. King turned to that occupation for the same reason as do other women: money.

Her father, a coal miner, would never have approved; he thought it wrong for women to work away from home at all. He also shared a commonly held miners' superstition: women bring bad luck in the mines.

When Mrs. King went to work at Westmoreland's Bullitt Mine, it was her first visit to a mine:

*I didn't know anything about a mine and I was surprised when I went down. I thought everything would be black and dreary, but when you get in and see all the white walls, it's not what you thought it was.**

Linda King had worked more than 5 years in sewing factories, earning a minimum wage:

There is no rest in a garment factory. When you go in and the buzzer goes at 7:30, you start work and you don't stop. It was a slave job. It was boring. Do you know that I worked with some women for 4 years and didn't even know their last names? Because you couldn't talk. No talking, none whatsoever, was allowed, even to the person sitting right next to you.

Now a roof bolter's helper, she earns nearly three times what she made in the garment factory. Factory workers earn a minimum wage of $2.90 an hour ($23.20 per 8-hour day), while the union scale for an underground mining trainee is $63.39 a day.

Some miners say that roof bolting is the most difficult job for a woman miner because it takes brute strength to bend the steel rods.

But you put it under the machine and then just step on it. Even though I've got the hardest job in the mine right now, I really don't work nearly as hard as in the sewing factory. Besides, the atmosphere is so different. You're not under as much pressure and tension as in the factory.

*The "white walls" are the result of the mine roof and walls being sprayed with ground limestone, commonly known as "rock dust." The rock dust helps cut down the level of coal dust in the mines and reduces the risk of explosion.

Mrs. King has a more important goal: "My daughter wants to go to college. She is 14 and wants to be a lawyer. I figured, if that's what she wants, I want to try to help her get it."

Sharon, West Virginia

Regina Fugate, 28, has been mining coal for the past 4 years:

I did it so I could have some time to myself. I was working four jobs before. I was a cashier at a local grocery store, plus I was cleaning three different places — a barber shop, a beauty shop, and an insurance office. As soon as I'd get done with all my jobs, I still would go home to an apartment with no furniture in it. So one day I put my application into the mine and then 6 months later I was a coal miner.

Now this woman miner has a new car, a completely furnished apartment, and all the clothes she needs. She can afford to buy things she wants and pay in cash.

Her first days in the deep mine unsettled her fellow workers. "They thought they'd have to watch their language. But pretty soon I was one of the workers, and now I don't think they even notice when I'm around."

Her fiancé, Jack, who is employed at a different mine, "thinks it's not right for me to be with all those men all day." But Miss Fugate would like to continue mining until she starts a family.

I won't be a miner forever, but the mines have gotten me on my feet as an individual and I think they'll keep me and my fiancé on our feet until we get a house finished. Mining has got me where I am and will get me where I want to go, and I feel real good about that.

Left: Linda King finds working as a roof bolter's helper at the Bullitt Mine in Big Stone Gap, Va., more challenging and better paying than her previous job in a garment factory. (Murray)

Right: Regina Fugate bought a new car and furnished her new apartment with earnings from a coal mine in Belle, W. Va. (Murray)

Working in Surface Mines

Judy Cordingly has a round smiling face and long, sandy hair that she tucks into a bun and then into her hardhat. Her work clothes are the regulation coal miner's boots, blue jeans, and a green checkered work shirt with a tee shirt underneath.

Mrs. Cordingly works as a blade operator, removing overburden in a large surface coal mine near Gillette, Wyo. She earns more than $11 an hour and works 5 days a week. She has been a surface miner for more than 2 years and for the first year and a half was the only woman working in production.

Surface mining was not difficult for her, she says, because she grew up on a farm operating heavy machinery. A desire to try something new, and a sister settled in Gillette, motivated her to move there. She brought her horses, which she planned to show, race, and breed. Because the horses were an expensive hobby, Mrs. Cordingly applied for work at the Carter Mining Company's Rawhide Mine, where the pay was higher than in most other jobs available to women.

She met her husband at the mine, where he operates a drag line. She was looking for a roping partner for a rodeo competition, and because he was the only cowboy she knew, offered to

pay his entry fee if he would ride with her. Today, they work together on the same crew, on the same shift.

Gary and Judy Cordingly live on an 800-acre ranch in Moorcroft, about 45 miles from the mine site. They have 40 head of cattle, raise wheat and hay, and train quarter horses.

Both would rather ranch than mine, but they need the money that they earn from coal mining to maintain their ranch.

Women Fought for Men's Benefits

Although women have always been active in the labor movement, their demands for fair wages and safe workplaces were traditionally in behalf of male miners. The history of the organizing of coal miners includes the efforts of generations of women, from the turn of the century when 1,000 women walked out of Wilkes-Barre (Pa.) Lace Mill demanding that the mill fire five women whose fathers and brothers were working in the mines during a strike, to the 1973-1974 Brookside strike in Harlan County, Ky., when the Brookside Women's Club showed up at the mine and in a rough confrontation persuaded the miners who were working to go home. A number of women were jailed for picketing at that mine and took their children to jail with them. Wives and widows were among the marchers in Washington who successfully sought passage of the Black Lung Benefits Reform Act of 1977.

Women's names dot the history of union organizing, names such as Mary Harris Jones — the legendary Mother Jones — who at the turn of the century organized women in behalf of miners. And Fannie Sellins, who was a 1911 UMWA organizer in the Pennsylvania coalfields. She went into the homes of coal miners as nurse and social worker, all the time preaching in behalf of the union. In 1919, she was killed by company guards.

And Josephine Roche. Daughter of wealthy parents, a graduate of Vassar College with a master's degree from Columbia University, she inherited several coal mines and became the president of the Rocky Mountain Fuel Company. When she took charge in 1927, one mine was on strike. In a confrontation between guards and strikers, 7 men were killed and 23 injured.

When she heard of the incident, she rushed to the mine and ordered everybody disarmed. The miners threw all the ammunition down a mine shaft and buried it. Josephine Roche then invited the UMWA to sign a contract with her company, which became the first Colorado company to recognize the union.

The *UMW Journal* noted: "She paid top wages and encouraged mine safety." Later she was instrumental in assisting John L. Lewis in establishing the UMWA Welfare and Retirement Fund and became its neutral trustee and director.

In the 1950s and 1960s, several women's organizations emerged in the coal regions, demanding better mine safety laws, improved accident prevention, and higher compensation for victims of Black Lung. "Women had a big hand in these organizations and in the improvements they eventually won," the *UMW Journal* reported.

For example, Francie Hager, known as Granny, joined by widows and disabled miners, formed the Appalachian Committee for Full Employment. The group became a "roving picket line," traveling from mine to mine asking the men to strike until working conditions improved. She became a leader in the Kentucky Black Lung Association and even when in frail health traveled wherever she was needed to fight for "coalfield justice."

Surface mining requires the ability to operate large equipment, a task Judy Cordingly found easy to master after growing up on a farm where she operated large machinery. (Wolford)

Overcoming Economic Fears and Superstitions

Women entering coal mining not only had to overcome economic worries on the part of male miners who feared losing jobs, but also had to surmount generations of superstition.

Resistance based on economic reasons was articulated by a legislator in Appalachia who reportedly said in 1976 that women were "going to take jobs away from the men and cause coal prices to go up because it will take two women to do the work of one man."

Some coal operators disagree. For example, Ira McKeever, president of the Colowyo Coal Company in Craig, Colo., who claims women "are better operators than men in terms of equipment maintenance, cleanliness, and less 'showoffishness.' They are very conscious of their position and are trying to do well." The company's goal is to employ up to 20 percent females.

R. C. Dizderich, mine manager of Utah International in Craig, also believes that "women are more stable employees" even though they require a little more training because most do not possess the skills and experiences that men have.

Women who work in deep mining face particular cultural biases ingrained in the lore of Appalachia. Not only did generations of miners believe that women brought bad luck into the mine, but many mountain people believe that coal mining is an improper occupation for women. Often those who work in the mines are accused of being immoral. In 1973, a minister was instrumental in organizing a march of miners' wives against a West Virginia mine that had hired women.

Composition of the Female Work Force

According to a 1975 UMWA study, its then 200 women members could be divided into two groups. One was made up of women in their mid-thirties, usually divorced with children to support; the other, of young, single women who intend to make mining a career.

The UMWA survey showed that more than 53 percent of these women provided the only support for their children. For those paying for day child care, the average cost was $30 a week; 10 percent of the women paid more than $50 per week. More than 42 percent of the married women miners said their husbands also worked in coal mining.

Rosemary McGilton, 29, a shuttle car driver in a North American Coal Company mine, lives in Willow Grove, Ohio, and is the main supporter of her family.

Married at 15 after quitting high school, she divorced 10 years later and found herself with two daughters, a son, and no marketable skills. She became a busboy, then waitress, then garment factory worker. She went to work in the mine as general inside laborer, a job that requires caring for conveyer belts, installing wooden posts that serve as temporary roof support, carrying cement blocks, cleaning track, and laying track. "I don't feel I should have to be married to provide for my children," she says.

First National Conference Is Held

In 1979, for the first time in the history of coal mining in this Nation, women combined to seek changes that affect them as a group.

The First National Conference of Women Coal Miners, held in July 1979 at West Virginia State College, drew more than 200 participants from eight States. They discussed issues that concern all coal miners, regardless of sex, such as health and safety, fair pay, job-bidding opportunities, disability benefits, and pension funds. The participants also dealt with specific issues that affect only women miners, such as job discrimination, child care, sexual harassment, and maternity benefits.

The law, regulations, and industry practice on these topics are changing, although not rapidly enough to satisfy some participants at the conference.

The Pregnancy Discrimination Act, passed April 29, 1979, states that pregnancy must be treated like any other temporary disability. The terms and coverage for such disability vary from policy to policy.

Child care poses a particular problem for single parents. Often no day care center facilities are available, and when they are, they are not open when the parent works on the evening or night shift.

Sexual harassment on the job was discussed at the conference. "Sexual harassment is a severe day-to-day problem for women miners," states Betty Jean Hall, legal counsel for CEP.

"The first women to go into a mine have a higher quit rate," she observes. "Those who enter 6 or 8 months later have it easier and stay longer, because many obstacles have been overcome."

These job-related concerns also were discussed at the UMWA Women Coal Miners Conference, held November 10, 1979, in Charleston, W. Va.

That conference was the first of its kind organized by the UMWA and was attended by about 70 female miners. It took place barely a month after Marilyn McCusker was killed in a roof fall in Coalport, Pa., the first woman miner killed in an underground mine.

In addition to speaking on issues, such as safety problems, that relate to all miners, the women emphasized problems of interest specifically to the female work force. They advocated research on the effects of coal mining on pregnant women and greater efforts to have properly fitting equipment.

Barbara Reagan, a West Virginia miner, said that she lost a finger because she did not have properly fitting heavy-duty work gloves. Proper boots and hardhats, for example, were mentioned as necessary equipment that is not available in women's sizes.

The women claimed they lacked training opportunities available to men, and therefore were relegated to more menial—and less well-paying—jobs. A woman from the West Virginia Human Rights Commission claimed that women seemed to reach higher paid classification 6 months later than their male counterparts.

Miners attending the conference asked the union to help establish day care and to combat hiring discrimination. There was also general agreement that companies should assume responsibility in disciplining those men who harass women on the job. "Sexual harassment leads to an intimidating situation . . . you have enough to deal with," commented Ms. Reagan, who also said that such harassment leads to safety hazards.

What Will the Future Bring?

Converging forces would seem to assure a steady rise in both the absolute number and the relative percentage of women in the coal-mining work force. First, the industry is expanding and, according to most indicators, will continue to expand in the 1980s; that means more jobs for more miners, men and women.

Second, a number of coal companies have pledged to hire women until 33 percent of their rosters are female. Third, women miners are showing increasing ability to work side-by-side with men in the mine, and to work effectively for their own interests outside the mine. They have organized themselves, selected forceful and articulate people to speak for them, retained legal counsel, and begun the long route of changing law, regulations, and industry practice.

The opening of coal mining to women could have the potential of substantially reducing the economic barriers that now separate men and women in regions where coal is dominant and where mining provides the only high-paying jobs.

Rosemary McGilton, drives a shuttle car at a mine near Willow Grove, Ohio, and supports three children. (North American Coal Co.)

Navajo Coal Miners

Tony Walker wears blue jeans and a turquoise ring, tends sheep evenings and weekends on his ancestral grazing land in Arizona, and lives quietly with his wife, Elaine, three children, and a grandmother. He is also part of a modern work force that produces approximately 23 million tons of fuel a year. Tony Walker is a Navajo coal miner.

Five and sometimes six days a week, he drives his pickup truck, with its "Arizona Coal is UMWA Coal" bumper sticker, to his job as a truck driver at the McKinley Mine, located on the eastern edge of the Navajo Reservation on the Arizona-New Mexico border.

McKinley Mine is one of four surface mines on the Reservation that together employ about 1,440 native miners and produce about 3 percent of the coal mined in the United States.

Total recoverable coal reserves for the Reservation are estimated at between 70 and 80 billion tons. Navajo coal fires generating plants located on the Reservation that supply electricity to portions of Arizona and New Mexico. Coal from the Black Mesa Mine is pumped by slurry pipeline to Nevada, where it fuels other electric generating plants.

Four mines operate on the Navajo Reservation. The Pittsburgh and Midway Coal Company's McKinley Mine outside Window Rock on the Arizona-New Mexico border employs about 240 Navajo miners. The largest operation is in the northern part of the Reservation 30 miles from Kayenta and 45 miles from the Utah border. Here, two Peabody Coal Company mines, Black Mesa and Kayenta, operate with about 750 Navajo miners. Navajo Mine, run by Utah International on the northeastern edge of the Reservation near Shiprock and Farmington, New Mexico, employs approximately 450 Navajo miners.

Coal mining on the Reserva-

tion is a relatively recent development. The first mine, McKinley, opened in 1959 and the three others followed within 15 years.

Major coal companies operate the mines on the Reservation under agreements with the Navajo Nation. The companies employ the miners, remove and transport coal, and pay a royalty to the Navajo Nation for each ton removed. A current controversy involves the amount of the royalty, which is substantially below the prevailing rate. The contracts in force today were negotiated several years ago on behalf of the Navajo Nation by the

Overleaf: Large coal mines and power plants opened on the Navajo Reservation in 1959. This plant in the Four Corners area of New Mexico powers homes in Southern California and Arizona. (Murray)

All four mines on the Reservation are surface operations; together, they produce about 23 million tons of coal annually. Top and right: Tony Walker, a Navajo miner, waited 2 years to become a dump truck driver at a mine near Window Rock, Ariz. (Mark Lennihan). Bottom left: Miners at three Navajo mines are UMWA members. (Wolford)

Bureau of Indian Affairs (BIA). Dissatisfied with BIA representation, the Tribal Council has won the right to negotiate on its own behalf.

Navajo miners' wages are comparable to those paid at surface mines in the East, the Midwest, and other parts of the West, although the Navajo may make more if he or she regularly works overtime and Saturdays. Figures from miners and coal companies on the Reservation show that most Navajo miners can earn more than $20,000 a year and that half earn more than $25,000 annually.

Coal companies operating on the Reservation must agree to abide by a policy of hiring qualified Indians on a preferential basis. The majority of miners working these mines are Navajos; members of other tribes as well as white and Hispanic-American miners also work in these mines. Nine Navajo women are employed as coal miners.

Many miners operate heavy equipment during the day — manipulating scoops on draglines 150 feet high or driving powerful bulldozers — then return to their native traditions and habits after work.

For some Navajo families, coal mining has helped to preserve ties. Simon Crank, a surface miner at Kayenta Mine, lives on a stretch of sun-dried range shared by sisters, mother, grandmother, and children. Four men in the family are coal miners — Simon Crank, his son, his brother, and his brother-in-law — and their combined annual salaries of about $100,000 support approximately 75 relatives.

Tony Walker's mining income also supports the Navajo way of life. He uses his increased salary, since leaving a job with a Comprehensive Employment Training Act program and becoming a coal miner, to buy horses, a horse trailer, and hay for his sheep.

Navajos, the Land, And Coal Mining

Reverence for land and the devouring process of surface mining collide on Simon Crank's land. He works in the cab of the giant scoop that removes the earth above the coal seam; he also herds sheep and cattle across his ancestral grazing lands, which lie 10 miles from Kayenta Mine. In the next year or two, he and his family may be relocated by the coal company that intends to mine the area.

When Simon Crank talks about his land, he tells of Father Sky cleansing Mother Earth, of wind and rain and soil blending and purifying each other.

Yet, the big machines of coal operations offer him a livelihood and a good chance to sustain his traditional ways with his family.

This cultural clash is repeated around other mines on the Reservation. Navajos struggle with their desire and need for the financial benefits of mining and their spiritual attachment to the lands that they have held for centuries.

For a variety of reasons, such as cultural beliefs about the sanctity of the land, concern about damage to the environment, or fear of air pollution from electric generating plants, some Navajos oppose surface mining on their Reservation. Coal development, they fear, will destroy their traditional way of life and bring industrial encroachment. They are particularly concerned about the potential aftereffects of surface mining, and contend that too little is done to reclaim the land for pastures and other uses after the mining machines move on. They are afraid they will lose their sheep, their grazing lands, and their open spaces.

Navajo miners earn the same wages paid at surface mines elsewhere. With overtime and Saturday work, they may earn more than $20,000 a year. Left: A miner washes up after work. Right: Alfred Halona (in hardhat), a miner for 17 years, with his family. (Murray)

The Miner as Navajo

The lifestyles of Navajo coal miners cannot be readily separated from other Indians on the Reservation. Some miners maintain a traditional Navajo lifestyle and spend their time and money on livestock, farming, and supporting large extended families that include grandmothers, aunts, cousins, and in-laws. Other coal miners opt for a more contemporary life of vacationing off the Reservation, buying pickup trucks and televisions, and finding entertainment in nearby border towns and cities.

Although two portable televisions perch on the shelves over the bed in their living room, the Walkers do not ordinarily spend evenings watching TV. Instead, they travel 15 miles to feed and tend the approximately 300 sheep they raise for wool. Livestock — sheep, cows, horses — are so fundamental to Navajo life that they are part of a daily ritual.

Calvin Segay, a miner who lives in a mobile home in Fort Defiance, Ariz., thinks that livestock is so essential to his family's stability that trailer life with no animals may cause his children to lose their Navajo values:

If you have livestock, you always have something to do almost every day. But no livestock —your kids start wandering out and it's kind of hard to bring them back home. What we got to do is keep the kids out of trouble by having something special to look forward to and care for, by learning how to appreciate life and respect life, so they will have a good understanding of what actually life means and how to treat life.

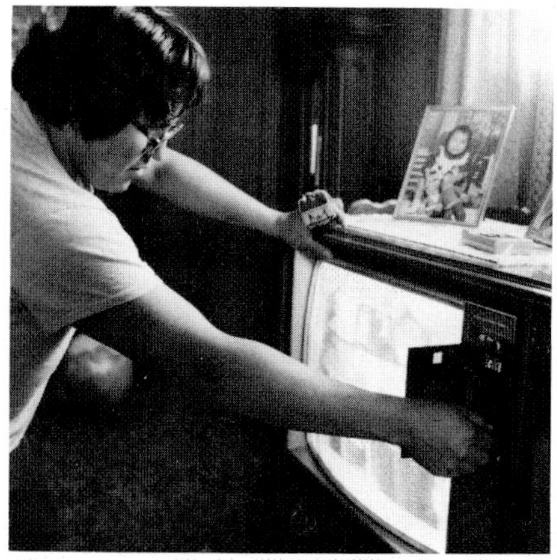

Some Navajos fear that coal mining will damage the land and that power plants will pollute the air. They prefer the traditional way of life. Other Navajos support coal mining, which provides miners with income for consumer items such as a television. (Murray)

Some Navajo mining families also farm. Elaine Walker spends part of her time growing corn on a 2-acre plot south of their home in St. Michaels.

She also does the family shopping. The "FedMart" in Window Rock is one of the few department stores on the Reservation and is stocked for the Navajo shopper who travels many miles, perhaps once a month, to load up on food and supplies. Miners and other Navajos buy work boots and wool jackets, garden tools and household furnishings, and groceries.

FedMart sells food mainly in bulk. Against the walls, lining the aisles, and stacked to the ceiling are steel crates jammed with 50-pound bags of dog food and 25-pound bags of flour. Elaine Walker spends about $100 a week on groceries. She shops when her husband joins a carpool to ride to work and leaves her the pickup truck.

Social activities among Navajos are frequently centered in the family. Around the Walkers are a grandmother next door, a sister down the hill, another sister living on the ridge above them, and an aunt who lives near the main road.

The Navajo Nation

The Navajo people have inhabited the Southwest for hundreds of years, living as hunters and herders in the redrock canyons, across the rugged deserts, and among the Ponderosa pines of the high country.

The Navajos' lands became part of the United States in 1848 when Mexico ceded the territory. The Treaty of 1850 officially made the Indians subject to the control and laws of this country. During the next 15 years, pioneers migrated westward and the Federal Government, fearful for the vulnerable settlers, decided

to rid the territory of Navajos and relocate them in a desolate area of eastern New Mexico. The relocation became the Long Walk, a tragic and unforgotten event in Navajo history; U.S. Calvary drove 8,000 Navajos on a 300-mile trek away from their homeland to an uninhabitable region where half of them would die in the next 4 years.

This devastation to the Navajo people prompted the Federal Government to negotiate the Treaty of 1868, a pivotal point in Navajo history. This treaty allowed the Navajos to return to their ancestral lands on a newly

The modern and the ancient live side-by-side on the Reservation. Left: Navajos load groceries at the FedMart in Window Rock, Ariz. Right: The traditional craft of rug weaving endures at Hubbell's Trading Post in Ganado, Ariz., and elsewhere on the Reservation. (Murray)

created 3.5-million-acre Reservation. It also mandated that the Navajos receive an education in English and be governed by the BIA.

The Navajos were awarded additional lands in the 1880s, and President Theodore Roosevelt increased their territory by ceding another 4 million acres in the early 1900s.

Navajos living on the Reservation are American citizens — they pay Federal income taxes, serve in the Armed Forces, and vote in national elections. They also are citizens of the Navajo Nation, a legal entity separate from any State and are subject to the Navajo Nation's criminal and civil laws. In 1923, the Navajo Tribal Council was elected as legal respresentative for all the Navajos who share in the ownership and use of the territory.

Today, the Tribal Council manages the affairs of the Reservation as an independent governmental body, subject to the approval of the BIA. The Council legislates and oversees, for example, police service, domestic relations, trade practices, and tribal ceremonies.

Housing Miners on the Reservation

Tony Walker has lived all his life in his present home, a four-room gray plaster and log building set on a dusty hillside facing a 6-foot stack of dead wood and scrubby land beyond. Unlike many homes on the Reservation, the Walkers' has electricity and running water. Its only source of heating, however, is a coal-and-wood-burning stove standing on the linoleum floor in the corner of the front room.

The home is amply furnished: throw rugs are scattered across the floor and blankets cover the couch. On the walls hang portrayals of Jesus and the Last Supper and family photographs of young men in military uniforms. The kitchen is clear of gadgetry and dishes. Only the square wooden eating table under the windows, empty counters extending from the sink, and white refrigerator catch the visitor's eye.

Although coal miners' earnings greatly exceed the average income on the Reservation, their salaries do not necessarily mean their housing is different from that of other Navajos. Where they live is usually determined by what is available and how they lived before becoming miners, not by what they can afford to buy or rent. The Living Conditions Survey shows that more than half the Navajo miners live in mobile homes; the next largest group, 38 percent, lives in single family homes, and the rest live in apartments and motels.

Housing deficiencies and shortages are a common condition on the Reservation. Of the approximately 24,500 housing units on the Reservation, 18,400 are substandard, lacking either water, electricity, or both. Officials estimate that 75 percent of the population does not have electricity. The traditional Navajo hogan — a circular one-room hut built from handhewn logs, stones, and earth — accounts for at least half the substandard housing on the Reservation.

Water, and with it indoor plumbing, is in short supply everywhere. The region is naturally arid, and laying pipes from any central source is expensive. Many miners haul water daily to their homes. The percent of Navajo Reservation homes with complete bathrooms, according to the Bureau of the Census, is 28 percent, compared with the national figure of 93 percent.

Some Navajo miners live near the mines on weekdays and travel home on weekends to be with their families. Top: Miners' homes in Kayenta, Ariz. (Murray) Bottom: Tony Walker's home. The wood fuels the stove, the only source of heat in the house. (Wolford)

Homes Are in Short Supply

Many Navajos encounter housing shortages. Newlyweds frequently live with parents, and grown children stay at home. Miners unable to find homes close to work may commute to their family home on weekends and live in motels during the week.

Dan Austin, a 52-year-old bulldozer driver at Kayenta Mine, sleeps in a company-owned trailer at the mine during the week and drives 45 miles on weekends to Chilchinbito, Ariz., to be with his wife and children.

Building new homes on the Reservation is a complicated and lengthy process. Navajos must obtain a homesite permit for their traditional grazing lands and then contract with a builder. In addition to securing the needed supplies and labor, the contractor must contend with Tribal, BIA, county, State, and Federal regulations.

A 25-year-old Navajo miner, for instance, left his job at Kayenta and moved to Fort Defiance to work at the McKinley Mine. He could not find a home for himself, his wife, and their three children, so he crowded them into a mobile home already occupied by his two sisters and their five children.

The People Are Scattered

A fundamental reason for the housing inadequacies is the inability of local government or private developers to provide basic services to a Nation of approximately 143,000 people who are spread out across the Reservation's 12.5 million acres of desert, forest, and red plains.

Navajos do not live in compact communities. Their "towns" may be a tiny grocery store and post office set back from a two-lane paved highway. Homes may be strung out along a dirt road, facing different directions, with no visible boundaries, or dispersed over the hills, their gray and earth-colored exteriors blending with the rocky, dusty terrain.

Educating Miners' Children

The Walkers have three children: Emmanuel, 5, Ophelia, 4, and "Baby Tony," 1 year old. The oldest attends St. Michael's Missionary School, a private elementary day school for grades one to three situated in the midst of a clump of trees just over the ridge from the Walker home. They pay $48 a month plus money for books and supplies for Emmanuel to attend St. Michael's, and both Tony and Elaine Walker believe that the expense is worth it.

Children of other miners and Navajos attend an assortment of schools: BIA schools; Arizona, New Mexico, or Utah public schools; private or mission schools; and contract schools run by communities. The choice between attending BIA or public school may depend on how close a child lives to a bus route. Contract schools are open in only five locales.

Children of families that roam with the livestock, driving the animals to the mountains for the summer and to the plains for the winter, may attend a variety of schools.

A miner from Old Saw Mill, a village 25 miles from Window Rock, recalls spending his youth "wandering back and forth" with his family's sheep between the mountains and the plains. He started high school at Fort Defiance Boarding School, switched to a public school in Flagstaff, transferred again to BIA's Fort Wingate Boarding School, and graduated from a public high school in Albuquerque.

Left: Navajo miner Simon Crank in his home. (Murray)

Right: Navajo children may attend BIA, public, contract, boarding, or missionary schools. Window Rock Elementary is a public school. (Berger)

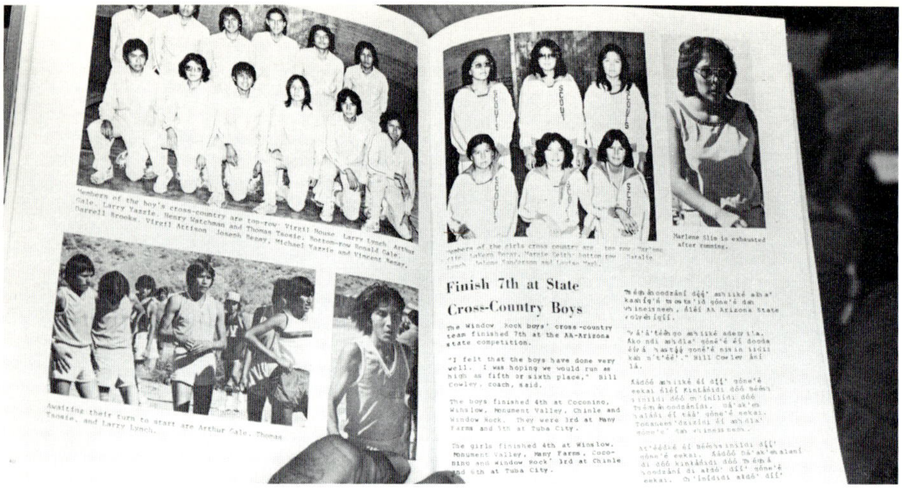

Window Rock School District
The Window Rock School District, headquartered in Fort Defiance, is comprised of an elementary, junior high, and high school combined into a central area of sandstone, red brick, and metal prefab buildings. Basketball courts, playgrounds enclosed by chain link fences, and weedy sidewalks vaguely separate the three schools. The high school was built in 1953 to hold 250 students — 1,000 use it today.

Children attending BIA and public schools usually have an opportunity to study Navajo culture and language. Dr. Kenneth Ross, the superintendent and himself an Indian, reports that up to one-third of entering children speak Navajo as a first language.

One miner who enrolled his son in public elementary school explains how he chose between English and bilingual instruction: "I chose bilingual because my boy would understand both ways. He will avoid confusion."

Despite the variety of schools serving the Reservation, educational attainment levels are low. Median educational achievement by adult Navajos, according to the 1970 Census, was 5th grade; the national median average was 12th grade. Coal miners on the Reservation reported in the Living Conditions Survey that only 58 percent finished high school, indicating a dropout rate of 42 percent — substantially higher than the national dropout rate of 26 percent.

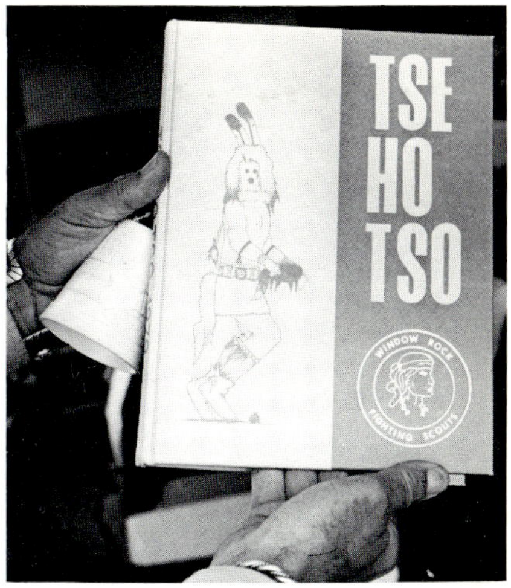

Schools on the Reservation combine traditional Navajo teaching with modern methods and customs. Top left and right: A high school yearbook, in both English and Navajo, is entitled "Tse Ho Tso," which in Navajo means "Fort Defiance." (Berger) Bottom left: Pupils at the Hunter Point Boarding School in Window Rock. (Murray) Center: Teenagers enjoy a junior-senior prom at the Academy of Kirkland near Farmington, N. Mex. (Murray)

Fort Defiance Hospital

Down the road from the Fort Defiance elementary school, tucked against a sandstone ridge, is the 76-bed Fort Defiance Memorial Hospital. When miners from McKinley Mine or their families need routine or emergency medical care, they come here.

The three-story, red brick building, with its healthy green front lawn, is set in the shade of tall trees. Constructed in 1936, the hospital looks from the outside as though it belongs on a college campus. Inside, that serenity is lost. The hospital is filled with people and most are Navajos.

In the family practice waiting room, some sit in chairs and wheelchairs, some lean against the walls, some have thermometers in their mouths — most are waiting to be seen by a physician.

The Fort Defiance Hospital has a staff of 14 physicians — family practitioners, general surgeon, obstetrician/gynecologist, pediatricians, and general medical officers — plus approximately 100 nurses, an optometrist, and a dentist. These health care professionals treat approximately 3,000 inpatients each year, and approximately 85,000 outpatients yearly.

The hospital administers only the most routine types of medical care; patients requiring sophisticated care are sent off the Reservation for treatment. In this way, Fort Defiance Hospital is similar to many of the other hospitals serving the Navajo Reservation.

Navajo miners may choose to visit private physicians and medical facilities off the Reservation because coal companies provide standard major medical insurance. A Navajo miner who lives in Fort Defiance explains:

I want to have the best doctor I can if I have an illness. When you go to an Indian Hospital you have to wait, and all they do, they just get as many people treated within 1 day, and they

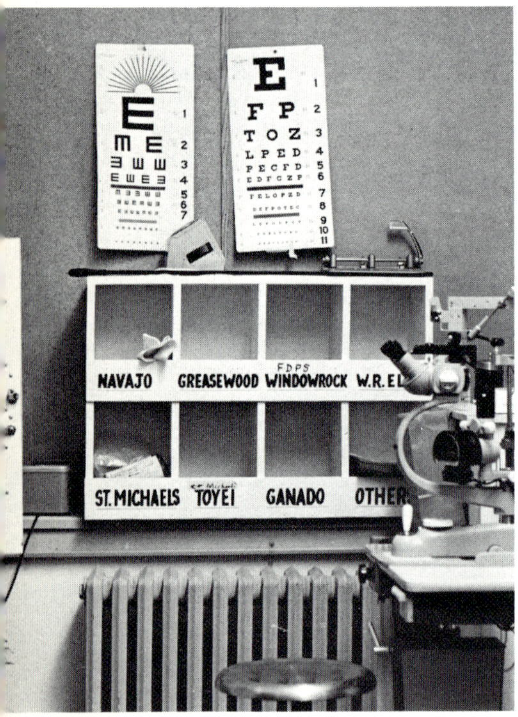

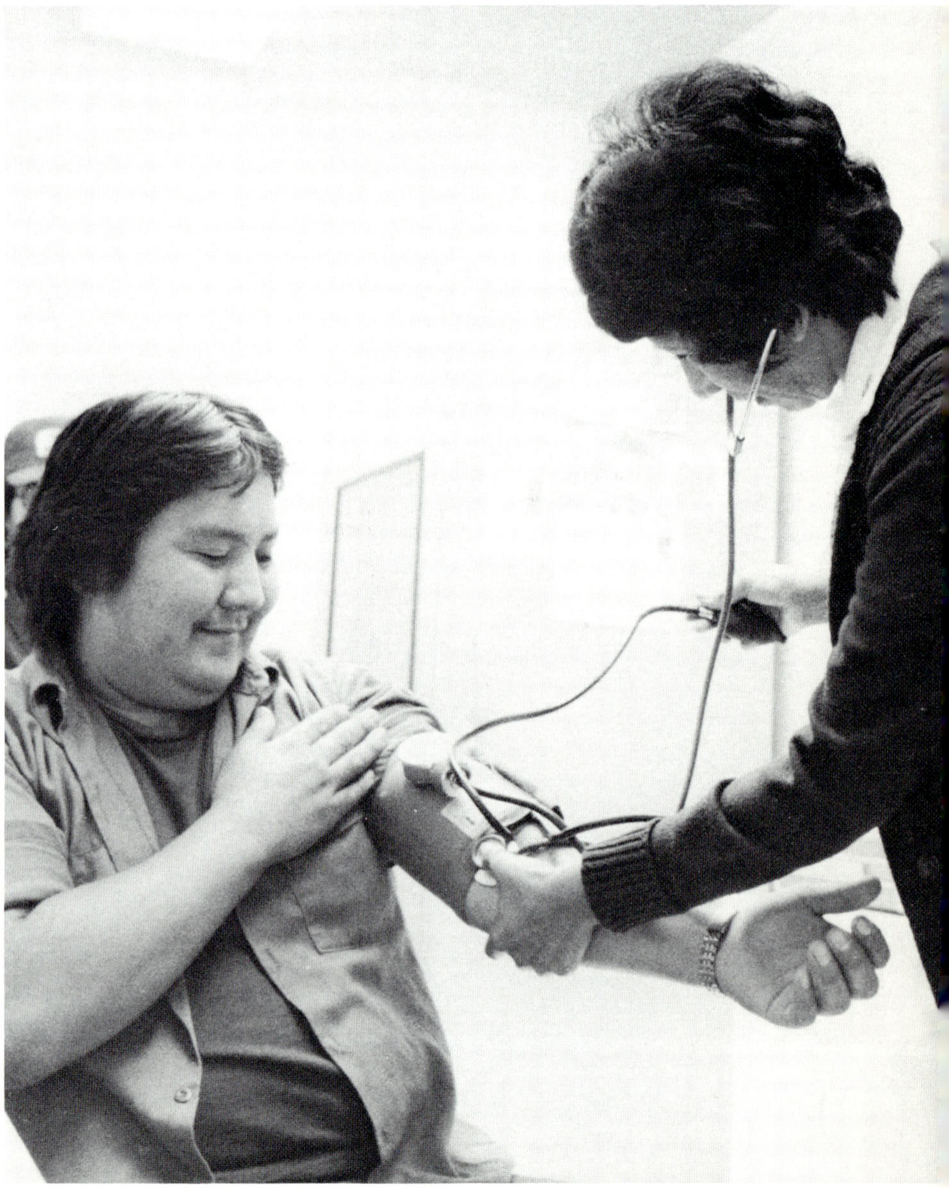

don't actually look. The only time that they get really serious is when a guy or a woman demands it and they are in critical condition, that's when they start really working on them. If I go to a private doctor I do a lot better than with an Indian doctor or a Government hospital.

Indian Health Service Treats Everyone

The health care that the Navajo miner and his or her family receives on the Reservation is provided by the Indian Health Service, which is part of the Department of Health, Education, and Welfare. These services are free, not just for the miner, but for all Navajos on the Reservation.

Health services are available at three types of facilities on the Reservation: hospitals (such as Fort Defiance Hospital), health centers, and health stations. Together they provide 58 locations for the treatment of medical problems.

Medicine Men Practice Today

Navajo medicine men have healed the sick of their tribe for centuries through soothing chants and herbal medicines. Today they continue to practice as traditional healers. Some miners say that they no longer go to medicine men for broken legs or injuries that can be better treated through conventional medical technology. They visit the medicine man for conditions such as headaches, stomach pains, and general irritability that are often caused not by infection but by emotional or family problems.

The medicine man studies for 10 to 15 years at a school on the Reservation. He is allowed to administer to patients in the hospital, and works in cooperation with the patient's physician. Medical insurance for some miners, depending on their union, pays for treatment by medicine men.

Navajo Health Is Below Average

Despite the array of health care facilities, Navajo health is generally worse than that of the population of the United States as a whole. Navajos die about 10 years younger than most Americans.

Illnesses that have become less prevalent in the United States as a whole have increased among the Navajos. The incidence of dysentery has risen by 189 percent over the past 10 years, hepatitis is up by 185 percent, and strep infections are up by 303 percent. The reasons for the increases, which include several infectious diseases associated with overcrowded conditions, appear to lie in the totality of life for some Navajos — the substandard housing, poor water, inadequate diet, and uneven health care.

Navajo miners obtain health care under major medical insurance programs offered by the coal companies. Left: Fort Defiance Hospital, part of the Indian Health Service, provides an array of health care services, including this eye clinic. (Berger) Center: A miner has his blood pressure checked. (Murray) Right: An Indian Health Service unit near the Fort Defiance Hospital. (Berger)

Coal Mining and the Future

The problems and hardships of the Navajos on the Reservation have been studied and documented since the territory was returned to them — more than 100 years ago — after the Long Walk.

Primitive housing, inadequate water supplies, few sanitation facilities, a bureaucratic school system, and ineffective health care facilities are chronic.

The introduction of coal mining into Navajo life has given some families added income to maintain their traditional ways. Coal, however, is a finite resource, even given widely differing estimates of the life expectancy of the Navajo mines. Some Navajos — and some outside observers — believe that coal mining may deplete this resource and leave the Navajos with an exhausted and empty land that lacks any other native industrial base.

Other Navajos — including the Tribal leadership — disagree with the notion that they will be exploited by coal development. They are seeking higher royalties from coal-mining leases.

They also have added the Navajo Nation to a potential energy cartel — the Council of Energy Resource Tribes (CERT) — made up of Indian groups representing Reservations in the United States that contain coal, oil, and uranium. In late 1979, CERT was fashioning itself along the lines of the Organization of Petroleum Exporting Countries (OPEC), the international cartel of oil-exporting nations.

Whose views will prove to be prophetic, however, is not yet known, and the story of the Navajo coal miners in 1979 must end with a question: To what extent will the rich coal deposits of the Reservation be converted into benefits for future generations of Navajos?

Navajos actively debate the proper course of future coal development on the Reservation. Questions arise about the environment, the survival of traditional ways of life, and who benefits from mining. For some Navajos, these questions remain unanswered. (Murray)

Appendix

The Living Conditions Survey

An important source of data for this report lies in the Living Conditions Survey, which involves a questionnaire administered to coal miners and interviews with coal miners and their families. The President's Commission on Coal undertook this survey in an effort to update existing data on miners and to obtain views of miners themselves on their community and living conditions.

The Commission developed a questionnaire to be used by the field survey teams to obtain information on miners' attitudes, beliefs, perceptions, and levels of satisfaction with respect to various aspects of their community and living conditions. In addition to seeking biographical and career-related data, the questionnaire was designed to elicit responses from miners regarding the following aspects of life in their communities: housing, water and sewage facilities, transportation, health services, shopping facilities and purchasing patterns, recreational and cultural opportunities, and educational facilities and services.

The questionnaire was submitted to and approved for use by the U.S. Office of Management and Budget (OMB). It was assigned OMB clearance number 44579011. A copy of the survey questionnaire follows this Appendix.

The tight project schedule did not permit a pretest of the questionnaire on a sample of coal miners. Although a pretest would have been desirable, the administration of the questionnaire in the field demonstrated that the survey items were understandable by and meaningful to the miners and, with only three exceptions, elicited unambiguous response data.

Selecting Field Survey Team Members

The field survey teams consisted of 15 persons selected on the basis of their experience with survey research projects, interviewing and interpersonal skills, writing ability, and motivation and stamina to maintain effective performance while on a rigorous work and travel schedule. No attrition occurred among members of the field survey teams.

Selecting the Sample

For purposes of the survey, the coalfields of the Nation were divided into five major regions: Northern Appalachia (Pennsylvania, Ohio, and northern West Virginia), Appalachia I (eastern Kentucky, southwestern Virginia), Appalachia II (southern West Virginia), Midwest/Alabama (Alabama, Illinois, Indiana, western Kentucky, Arkansas, and Oklahoma), and West (Colorado, New Mexico, Arizona, Utah, Wyoming, and Montana).

Within the five regions, 29 population centers were chosen as the focus of primary survey team activity. The number of population centers within each region varied from three to eight depending upon the concentration of coalmining employment in each region.

The Commission asked the Bituminous Coal Operators' Association (BCOA), the United Mine Workers of America (UMWA), and the National Independent Coal Operators of America (NICOA) to provide lists of mines within 60 miles or a 2-hour drive (whichever was more distant) of each population center. The mines were divided between surface and underground mines based on the proportion of employment by type of mining in the area. To provide

Left: Jasper, Ala. (Murray)

the largest possible universe of miners from which random selections might be made, larger mines (employment of 200 was used as a guideline) were chosen within each of the population centers. The Commission staff added several independent mines that fit the same criteria.

The final criterion for selection of mines was broad geographical distribution within each of the population centers. This was done to assure that the full range of living conditions throughout the population center would be represented.

This process resulted in the selection of 55 mines to be visited. At each mine, team members identified the miners and asked them to fill out a questionnaire and to participate in group interviews. Each miner was randomly selected according to the next-to-last digit of his or her Social Security number. Anonymity was guaranteed to all who filled out questionnaires and agreed to group interviews. Team members then asked a small number of miners from each mine to participate in at-home interviews with their families. Survey teams tried to select miners for home interviews who were representative of the work force. Miners had the option to refuse to participate, but only a few did.

For the at-home interviews, survey teams asked the participants to sign a release form allowing the Commission to use their names and direct quotes in this report. Only a few miners objected to the release.

Administering the Field Survey
Survey team members administered the questionnaire at the mines before or after a shift. Miners were selected from all shifts. Thus, some interviews were conducted shortly before the "hoot owl" shift began at midnight, and others at dawn as well as during the day.

Following a briefing by the team members, miners completed the questionnaires. On the Navajo Reservation, team members employed a native interpreter during questionnaire administration and for interviews.

Team members conducted the group interviews immediately following administration of the questionnaire. Team members tape recorded the interviews and used quotes, without attribution, in their field reports to Commission headquarters in Washington, D.C. Although miners were told that they could speak off the record, with the tape recorder turned off, none did so. On the contrary, miners often insisted that their views be put "on the record." Team members guided the major portion of the discussion, centering on issues dealt with in the questionnaire; at the end, miners could raise any point they desired.

Cooperation from miners in agreeing to at-home interviews was excellent. Field teams were able to conduct interviews in 189 homes, representing 15 percent of the total sample of 1,299 miners who responded to the questionnaire.

Interviewers both tape recorded and kept notes on the at-home interviews. Often a miner's spouse would participate, and sometimes children, grandparents, and other family members contributed to the discussion. Interviewers also made notes on the conditions of the home, the neighborhood, and the roads leading to the miner's home.

Analyzing the Data

The Commission received all completed questionnaires from the field teams by early June 1979, and staff checked them for completeness of response. Questionnaire responses were then entered into a computer data file, which permitted cross-tabulation correlations and determination of means on selected questions.

The field survey teams prepared rough field reports based upon the interviews and telephoned them to Commission headquarters. When they returned to Washington, the team members entered into 2 days of intensive debriefings by the Commission staff on all aspects of their field work; they discussed their regions and they participated in group sessions to compare impressions among regions.

Validity of the Sample
The Commission staff compared the national results of the questionnaire survey with known national data on coal miners. The survey sample consisted of 1,299 respondents, which was 0.6 percent of the coal miners in the United States in the spring of 1979. The Commission's survey* data were compared with other known data on the work force to test whether the sample was representative of known age and geographical distribution within the work force.

The questionnaire sample was found to be representative in terms of age distribution. The actuarial tables (dated June 30, 1978) of the UMWA Health and Retirement Funds were used as a proxy for a national age profile. The percentages of respondents to the Living Conditions Survey run consistently above those of the UMWA figures for miners up to the age of 30; they agree with the UMWA figures for miners aged 30 to 45; and they run consistently below those of the UMWA for miners more than 45 years of age. These differences may be attributed to the fact that

*Eighteen questionnaires were invalidated as incomplete.

the tables do not include western miners. It is generally acknowledged that there are a larger number of young miners in the western coalfields. This difference would account for larger percentages in the younger age groups in the Living Conditions Survey sample. (See Table 1.)

The sample is representative of employment figures in three of the five regions. The northern Appalachian region was underrepresented and the western region was over-sampled. Commission staff decided to increase the sample for the West because of the variety of living conditions found in that region. Conversely, the northern Appalachian region possesses fewer diversified living conditions and thus, was sampled less than employment figures would indicate. (See Table 2.)

The sample was not tested on the basis of type of mining or of union or nonunion status, although the miners in the samples represent all these factors. These factors were not used as a basis for testing because living conditions may be regarded as independent of them.

The guidelines (employment of 200) used for selection of mines within each of the population centers chosen for study inevitably biased the sample toward those employed in larger mines. Commission staff believes that this approach had little effect on the results obtained in the study.

Table 1.
Age Distribution of U.S. Coal Miners

Age of Respondent	United Mine Workers of America Health and Retirement Funds Actuarial Table[1]	Living Conditions Survey
19 and under	0.43%	1.11%
20-24	14.78%	16.88%
25-29	20.56%	24.20%
30-34	16.13%	16.24%
35-39	11.96%	11.31%
40-44	9.12%	9.24%
45-49	7.47%	6.45%
50-54	8.19%	6.69%
55-59	7.04%	5.18%
60-64	3.80%	2.39%
65 and over	0.52%	0.32%

[1]*Distribution of active miners covered by the UMWA Health and Retirement Funds as of April 1, 1978. Data do not include miners in the western coalfields.*

Table 2.
Geographic Distribution of Active Coal Miners

Region	Actual Work Force, as Compiled by the Mine Safety and Health Administration January-June 1979[1]	Distribution of Living Conditions Survey Respondents
Northern Appalachia	30.0%	18.0%
Appalachia I	22.0%	19.0%
Appalachia II	18.0%	24.0%
Midwest/Alabama	22.0%	22.0%
West	8.0%	17.0%

[1]*Percentage of total national work force, based on average employment in coal mines.*

Critique of the Questionnaire Data

The questionnaire was designed to provide a qualitative study of the attitudes and perceptions held by miners and members of their families toward various facets of their communities. It provided information that could then be explained in greater depth in the group and at-home interviews. The questionnaire was not designed to yield "hard" data that could be subjected to rigorous statistical analysis. The reader should keep this fact in mind when reviewing the appended questionnaire and when reading this document.

When questionnaire responses were compared with data collected from the miner group interviews and the at-home interviews, certain inconsistencies were noted with respect to the intensity and direction of attitudes and satisfaction levels expressed. For example, the vast majority (82 percent) of the miners who completed the questionnaire indicated that they were "very satisfied" or "satisfied" with their housing. A large majority of these miners also indicated that they were satisfied with the water system (78 percent) and the sewer system (83 percent) that served their homes.

Yet, when the results of the group and at-home interviews were analyzed, the level of satisfaction expressed with these and other facets of living conditions seemed to be at variance with the questionnaire responses. Inasmuch as these interviews were designed to elicit more detailed nuances of, and explanations for, attitudes and perceptions expressed on the questionnaire, this finding appears to be natural and predictable. A miner may well be satisfied overall with the quality of his or her housing, but, when questioned in depth, perceives certain aspects of community living as unsatisfactory.

Other specific problems arose in the administration of the questionnaire and analysis of data derived from it:

Data from four questions (#9, #10, #11, and #34) could not be included in the questionnaire data analysis because of either the large number of no responses or invalid responses. The name of county and State of birth (#9) as well as county and State of present residence (#10) were asked as a means of measuring mobility or migration patterns of miners.

The inability to identify questionnaire responses by job classification of miners (#11) poses no significant limitation, inasmuch as this is not a study of working conditions or labor-management relations.

The item regarding ownership of property (#34) other than a home was apparently ambiguous and yielded such a diversity of responses that the data were useless.

Data from questionnaire item (#46) regarding the accessibility and use of various modes of transportation (air, bus, or train service, and "other"), could not be used because of the high nonresponse rate. This item appeared to be unambiguous, and no clear rationale for the high nonresponse rate can be offered.

Although all the questionnaire items #63 through #69 had a high nonresponse rate, all of these items — with the exception of item #65 — yielded sufficient data to warrant inclusion in the report. The data from item #65, regarding availability in the community of vocational-technical schools, adult education, a community college, or a 4-year college, were so incomplete as to necessitate their elimination from the data analysis.

The loss of these questions from the data, while not significant, precluded data analysis of topics that they were designed to measure.

THE PRESIDENT'S COMMISSION ON COAL

SURVEY OF COMMUNITY CONDITIONS
IN COAL PRODUCING AREAS

Instructions

On this questionnaire, you are asked a series of questions about yourself and your family; your experience as a coal miner; your housing; the water and sewer systems serving your home; transportation and road conditions; health services, shopping facilities, and recreational activities available in your community; and other information about the area in which you live. On some questions you will check a "yes" or a "no". On other questions you will check a choice of answers which is listed, choosing the one answer which is right for you. On some questions, you might be asked to check more than one answer if more than one applies to you. On some questions are you asked to check a rating. On some questions you are asked to write in a brief answer. Instructions in the brackets with each question tell you how to mark that question. Please answer every item in the questionnaire.

This report is authorized by E.O. 12103. Your voluntary cooperation is needed to make the results of this survey comprehensive, accurate and timely.

The information collected on this form by The President's Commission on Coal will be held in confidence and used for statistical purposes only.

Form Approved: OMB No. 44579011

(Average or aggregate figures for 1,281 respondents are included where feasible.)

1. What is your age? 34.9 yrs.
2. What is your sex?
   96.9% Male 3.1% Female
3. What is your marital status? (please check one)
   10.3% Single 71.7% Married 17.3% Divorced 0.7% Widowed
4. How many children do you have living at home with you? 2.2
5 How many family members do you support? 3.4
   3.3 in your house 1.9 outside your home
6. What do you think your total income will be in 1979? $19,442.33
7. Does your wife (husband) work? (please check one)
   26.5% Yes 73.5% No
   If "Yes," about how much money will she (he) earn in 1979? $9,314.51
8. What is the highest school grade you have completed? 11.4 mo.
9. Name of county and state in which you were born:
10. Name of county and state in which you now live:
11. What is your job classification?
12. What shift do you work?
13. How long have you been a coal miner? 9.7 years
14. How long have you worked for this company? 7.3 years
15. How long have you worked in this mine? 6.3 years

-2-

16. How long have you worked at your present job? 4.1 years

17. In your family, who is or has been a miner? (please check any that apply)

a. children 6.6%
b. brothers 38.6%
c. sisters 2.3%
d. father 55.5%
e. grandfather 36.5%
f. great grandfather 12.3%
g. none 17.3%

18. Do you plan to stay in coal mining until you retire? (please check one)

74.1% Yes 8.5% No 17.4% Don't Know

19. Would you like to have your child (or children) become a coal miner? (please check one)

23.3% Yes 53.7% No 22.9% Don't Know

20. Were you laid off from work in the last year? (please check one)

17.9% Yes 82.1% No

If "Yes," for how long? 10.8 weeks

21. Were you placed on short work weeks in the last year? (please check one)

30.2% Yes 69.8% No

If "yes," for how long? 13.6 weeks

22. Have you worked overtime during the past year? (please check one)

81.6% Yes 18.4% No

If "yes," please estimate the amount of overtime. 9.6 hrs. per week

23. Do you live in a: (please check one)

a. single-family house 68.2%
b. apartment 5.2%
c. condominium/townhouse 1.0%
d. mobile home 23.5%
e. other 2.0%

________________ (indicate)

-3-

24. Do you own your home? (please check one)

74.4% Yes 25.6% No

If "yes" do you : (please check one)

a. own outright 34.1%
b. have a mortgage 29.8%
c. make payments 36.0%

25. Do you rent your home? (please check one)

19.6% Yes 80.4% No

If "yes," do you rent from: (please check one)

a. coal company 0.5%
b. individual 87.8%
c. realty company 8.9%

26. Do you live with relatives? (please check one)

7.0% Yes 93.0% No

If "yes," are they your relatives? 88.1% Yes 11.9% No
Are they your spouse's relatives? 11.9% Yes 88.1% No

27. How much money do you spend each month for housing (mortgage, rent or payments)?

$ 241.38

28. Is suitable housing, that you can afford to buy available to you in your area? (please check one)

38.5% Yes 48.3% No 13.2% Don't Know

29. Is suitable rental property, that you can afford, available to you in your area? (please check one)

34.2% Yes 47.9% No 17.9% Don't Know

30. Is affordable land, for building homes on, available to you in your area? (please check one)

36.6% Yes 52.9% No 10.5% Don't Know

-4-

31. How satisfied are you with your present housing? (please check one rating)

a. 31.7% very satisfied
b. 49.7% satisfied
c. 14.1% dissatisfied
d. 4.5% very dissatisfied

32. How do you think your housing compares with that of other people in your community? (please check one rating)

a. 19.3% better than most
b. 75.0% same as most
c. 5.7% worse than most

33. How do you think your housing compares with that of most Americans? (please check one rating)

a. 13.0% better than most
b. 69.4% same as most
c. 17.6% worse than most

34. Do you own property other than a home? (please check one)

24.7% Yes 75.3% No

If "yes" specify: ________________

35. What type of drinking water system serves your dwelling? (please check one)

a. municipally owned public water system 55.6%
b. privately owned public water system 11.6%
c. private well 30.5%
d. other

________________ 2.3% (indicate)

36. How satisfied are you with your present water system? (please check one rating)

a. 27.6% very satisfied
b. 47.3% satisfied
c. 15.3% dissatisfied
d. 9.9% very dissatisfied

-5-

37. How do you think the water system that serves your dwelling compares with that of most Americans? (please check one rating)

a. 26.7% better than most
b. 53.1% same as most
c. 20.2% worse than most

38. What type of sewage system serves your dwelling? (please check one)

a. public sewer 39.3%
b. septic tank 50.8%
c. cesspool 5.5%
d. other 4.5%

39. How satisfied are you with the sewage system that serves your dwelling? (please check one rating)

a. 22.6% very satisfied
b. 62.5% satisfied
c. 9.0% dissatisfied
d. 5.9% very dissatisfied

40. How do you think the sewage system which serves your dwelling compares with that of most Americans? (please check one rating)

a. 14.2% better than most
b. 70.9% same as most
c. 14.9% worse than most

41. Do you own a:

	Yes	No	How many
car	97.8%	2.2%	1.3
truck	96.9%	3.1%	1.1
other (specify)	8.7%		1.1

42. How long does it take you to get to work? 29.5 minutes

43. What is the condition of the roads you travel to work? (please check one rating)

a. 8.3% good
b. 29.1% average
c. 32.5% poor
d. 30.1% very poor

44. How satisfied are you with the condition of the roads in your area? (please check one rating)

a. 2.7% very satisfied
b. 18.1% satisfied
c. 36.5% dissatisfied
d. 42.7% very dissatisfied

45. How do you think the roads in your community compare with roads in other American communities? (please check one rating)

a. 3.3% better than most
b. 28.4% same as most
c. 68.3% worse than most

46. Are the following forms of transportation accessible to you?

	Yes	No
a. air service	55.0%	45.0%
b. bus service (excluding school bus)	59.5%	40.5%
c. passenger train service	23.4%	76.6%
d. other	24.5%	75.5%

If accessible, do you use:	Yes	No
a. air service	27.3%	72.7%
b. bus service (excluding school bus)	19.8%	80.2%
c. passenger rail service	9.2%	90.8%
d. other	28.1%	71.9%

47. How long does it take you to drive from your home to your doctor's office or clinic? 23.7 minutes

48. How long does it take you to drive from your home to the nearest hospital? 25.7 minutes

49. How satisfied are you with the health care facilities now available to you and your family? (please check one rating)

a. 11.0% very satisfied
b. 55.3% satisfied
c. 22.7% dissatisfied
d. 11.0% very dissatisfied

50. How do you think the health care facilities available to you compare with those available to most Americans? (please check one rating)

a. 11.3% better than most
b. 60.2% same as most
c. 28.5% worse than most

51. Please check the services available to you:

	Yes	No
a. ambulance service	92.4%	7.6%
b. drugstores	89.1%	10.9%
c. dentists	88.4%	11.6%
d. eye doctors	82.4%	17.6%
e. fire and rescue service	80.9%	19.1%
f. trained para-medics	52.1%	47.9%
g. social service agencies	60.0%	40.0%

52. Please check the items you own:

a. television 93.2%
b. dishwasher 34.6%
c. clothes washer 84.5%
d. clothes dryer 80.6%
e. refrigerator 90.1%
f. stereo 80.6%
g. camper or trailer 17.3%
h. motor home 4.1%
i. motor boat 13.0%

53. How long does it take you to get to your usual grocery store?

17.0 minutes

54. Please check the types of stores available to you:

a. shopping mall 54.4%
b. fast-food shops 67.4%
c. discount stores 67.1%
d. chain stores 64.2%
e. supermarkets 86.7%
f. department stores 71.3%
g. speciality stores 49.2%

55. How long does it take you to get to a large shopping center?

48.8 minutes

56. How often do you or your family shop in large shopping centers? (please check one)

a. never 2.5%
b. less than once a month 27.6%
c. once a month 20.5%
d. twice a month 23.2%
e. once a week 21.6%
f. more than once a week 4.5%

57. How satisfied are you with the stores that are available to you in your area? (please check one rating)

a. 12.6% very satisfied
b. 63.7% satisfied
c. 18.5% dissatisfied
d. 5.3% very dissatisfied

58. How do you think the stores available to you compare with the stores available to most Americans? (please check one rating)

a. 7.0% better than most
b. 68.9% same as most
c. 24.0% worse than most

59. Are the following available in your area? (please check all that apply)

74.2% movie theater
85.3% restaurants
70.1% parks
63.4% bowling alleys
77.6% library
31.4% plays or concerts
82.4% hunting and fishing
69.7% recreation facilities (basketball courts, baseball fields, etc.)
66.4% swimming pools
34.9% youth centers
39.0% community centers
64.1% playgrounds

60. How satisfied are you with the recreational facilities available to you? (please check one rating)

a. 8.2% very satisfied
b. 53.5% satisfied
c. 25.3% dissatisfied
d. 13.0% very dissatisfied

61. Do you usually travel away from home while on vacation?

76.4% Yes 23.6% No

If "yes" how many miles do you usually travel on vacation? 752.5 miles

62. How many children do you have in

- elementary school 1.6
- junior high school 1.2
- high school 1.4
- community college 1.0
- vocational/technical school 7.9
- four year college or university 1.2

63. How long does it usually take your children to get to school?

21.3 minutes

64. Are you satisfied with the transportation your children use to get to school? (please check one)

82.1% Yes 17.9% No

65. Are the following available to you in your community:

	Yes	No
a. vocational/technical school	75.2%	24.8%
b. adult education	75.1%	24.9%
c. community college	66.7%	33.3%
d. four year college or university	49.9%	50.1%

66. How satisfied are you with the quality of education available to you and other adults in your community? (please check one rating)

a. 7.3% very satisfied
b. 59.6% satisfied
c. 25.5% dissatisfied
d. 7.5% very dissatisfied

68. How satisfied are you with the quality of education available to children in your community? (please check one rating)

a. 9.9% very satisfied
b. 62.2% satisfied
c. 19.4% dissatisfied
d. 8.5% very dissatisfied

69. How do you think the quality of education available to children in your community compares to that available to most American children? (please check one rating)

a. 12.4% better than most
b. 64.7% same as most
c. 22.8% worse than most

-10-

70. Has people's pride in your community: (please check one)

a. increased in the last 5 years? 27.9%
b. decreased in the last 5 years? 23.4%
c. stayed about the same? 48.7%

71. Do you feel that the future of the coal mining industry will continue to provide you with a job? (please check one)

63.2% Yes 13.3% No 23.5% Don't Know

72. Are you satisfied with your decision to be a coal miner? (please check one)

79.5% Yes 14.9% No 5.6% Don't Know

73. Would you make that decision again? (please check one)

73.2% Yes 18.3% No 8.5% Don't Know

74. Do you think that the American public has an accurate picture of how coal miners work and live? (please check one)

10.3% Yes 89.7% No

If "No," do you feel the American public thinks that: (please check one)

a. Things are better for coal miners than they actually are? 71.1%
b. Things are worse for coal miners than they actually are? 28.9%

THANK YOU AGAIN FOR YOUR COOPERATION IN COMPLETING THIS QUESTIONNAIRE.

References

The following references proved to be the most useful in researching the background for this report. That research also included review of a large number of published reports, and made use of information developed by the Commission in other areas of its overall effort. This list of references is intended to be of help to the general reader, and does not constitute a complete bibliography of current literature in the field.

Academy for Contemporary Problems. *Living Conditions in the U.S. Coalfields: Three Decades of Change.* 3 vols. Columbus, Ohio: 1979.

American Medical Association. *Distribution of Physicians in the United States 1970.* Chicago: American Medical Association, 1971.

American Medical Association. *1978 Directory of Residency—1979 Training Programs: Accredited by the Liaison Committee on Graduate Medical Education.* Chicago: American Medical Association, 1978.

American Medical Association. *Physician Distribution and Medical Licensure in the U.S., 1977.* Chicago: American Medical Association, 1977.

Appalachian Alliance. *Appalachia 1978: A Protest from the Colony.* 1978.

Appalachian Regional Commission. *Appalachia — A Reference Book.* Washington, D.C.: Appalachian Regional Commission, June 1977.

Appalachian Regional Commission. *Appalachian Children and Their Families; A Statistical Profile.* Prepared for the Conference on Raising a New Generation in Appalachia, November 14-17, 1978. Washington, D.C.: Appalachian Regional Commission, 1978.

Armbrister, Trevor. *Act of Vengeance: The Yablonski Murders and Their Solution.* New York: Saturday Review Press/E. P. Dutton & Co., Inc., 1975.

Axelrod, Jim. *Housing Problems in the Eastern Coalfields* (appendix). In Lewis, Helen, and Associates. *Coal Productivity and Community: The Impact of the National Energy Plan in the Eastern Coalfields.* Prepared for the U.S. Energy Department, Washington, D.C., February 1978.

Barnum, Darold T. *The Negro in the Bituminous Coal Mining Industry,* Report No. 14. In *The Racial Policies of American Industry.* Philadelphia: Industrial Research Unit, Department of Industry, Wharton School of Finance and Commerce, University of Pennsylvania, 1970.

Carter, Richard. *The Doctor Business.* New York: Doubleday & Company, 1968.

Caudill, Harry M. *Night Comes to the Cumberlands.* Boston and Toronto: Little, Brown and Company, 1962.

Center for Community Change. *The Lifeblood of Housing: A Report to the Commonwealth of Pennsylvania on Housing Investment and Conservation.* Prepared under a grant from the Appalachian Regional Commission. Alexandria, Virginia: Mt. Vernon Publishing Co., 1976.

Coal Employment Project. *Women in the Coal Industry.* Oak Ridge, Tennessee: 1978.

Collis, Edgar L. "The Coal Miner: His Health, Diseases, and General Welfare." *The Journal of Industrial Hygiene* VII-5 (1925): 221-243.

CONSAD Research Corporation. *The Basic Eastern Coalfield Housing Situation.* Paper prepared for the President's Commission on Coal. Pittsburgh: 1979.

Council on Environmental Quality. *Ninth Annual Environmental Quality Report.* Washington, D.C.: Government Printing Office, 1978.

Federal Writers' Project of the Work Projects Administration in the State of Kentucky. *Kentucky: A Guide to the Bluegrass State.* American Guide Series, sponsored by the University of Kentucky. New York: Hastings House, 1939.

Federal Writers' Project of the Work Projects Administration in the State of Pennsylvania. *Pennsylvania: A Guide to the Keystone State.* American Guide Series, cosponsored by the Pennsylvania Historical Commission and the University of Pennsylvania. New York: Oxford University Press, 1940.

Federal Writers' Project of the Work Projects Administration in the State of West Virginia. *West Virginia: A Guide to the Mountain State.* American Guide Series, sponsored by the Conservation Commission of West Virginia. New York. Oxford University Press, 1941.

Guenther, Sue; Spangler, Don; and Given, Lynda. *Coping with Growth: Energy Impacts in Five Appalachian Counties.* New York: National Association of Counties Research, Inc., 1979.

Gunther, John. *Inside U.S.A.* New York and London: Harper & Brothers, 1947.

Hull, Arthur M., and Hale, Sydney A. (eds.). *Coal Men of America.* Chicago: The Retail Coalman, Inc., 1918.

Joint White House/West Virginia Governor's Office Project. *Report of the Working Group on Mountainous Housing in West Virginia.* Charleston, West Virginia: Governor's Press Office, 1978.

Kerr, Lorin E. "Black Lung: A Case Study." Duplicated. United Mine Workers of America.

Landis, J.R.M. "The Relation of Organic Dust to Pneumonokoniosis." *The Journal of Industrial Hygiene* VII-I (1925): 1-5.

Lindbergh, Kristina, and Provorse, Barry. *Coal: A Contemporary Energy Story.* Edited by Robert Conte. Seattle: Scribe Publishing Corp., 1977.

Maurice, Johanna. "Why Did Mine Workers' Health Funds Fail?" *Charleston Daily Mail.* Charleston, West Virginia: 23 November 1968.

Metz, William C. *Residential Aspects of Coal Development.* Unpublished paper presented at the American Institute of Planners Conference, October 10-12, 1977.

Miller, Tom D. "Who Owns West Virginia?" Reprinted from *The Herald-Advertiser and The Herald-Dispatch.* Huntington, West Virginia: The Huntington Publishing Co., 1974.

Morgan, W. Keith C.; Burgess, Dean B.; Jacobson, George; O'Brien, Richard J.; Pendergrass, Eugene P.; Reger, Robert B.; and Shoub, Earle P. "The Prevalence of Coal Workers' Pneumonoconiosis in U.S. Coal Miners." *Archives of Environmental Health* 27 (1973): 221-226.

Mountain Life and Work (magazine). January 1979.

National Transportation Policy Study Commission. Draft report and other materials of the Task Force on Coal Hauling. Washington, D.C.: 1979.

Old West Regional Commission. *Construction Worker Profile.* Washington, D.C.: 1977.

Out of the Dark 2: Mining Folk. Barnesboro, Pennsylvania: Northern Cambria High School, 1977.

Rhodenbaugh, Suzanne. "Coalfield Clinics Strugle to Survive After UMW Contract Changes." *The Mountain Eagle.* Whitesburg, Kentucky: Jan. 4, 1979.

Rosen, George. *The History of Miners' Diseases.* New York: Shuman's, 1943.

Ross, M.H. "The Appalachian Coal Miner: His Way of Living, Working, and Relating to Others." *Annals of the New York Academy of Sciences* 200 (1972): 184-185.

Rural America (Newsletter). Washington, D.C.: November 1978.

Sheppard, Muriel (Earley). *Cloud by Day; Story of Coal and Coke and People.* Chapel Hill: University of North Carolina Press, 1947.

Skidmore, Hubert. *Hawk's Nest.* New York: Doubleday, Doran and Company, Inc., 1941.

Taubenhaus, Margorie, and Penchansky, Roy. "The Medical Care Program of the United Mine Workers Welfare and Retirement Fund." In *Health Services Administration: Policy Cases and the Case Method.* Edited by Roy Penchansky. Cambridge, Massachusetts: Harvard University Press, 1968.

UMW Journal. Various years.

Urban Studies Center, University of Louisville. *Housing in Kentucky: The 1978 Annual Housing Report.* Louisville: 1979.

Urban Studies Center, University of Louisville. *Housing Needs Analysis: Commonwealth of Kentucky,* Update 1977. Louisville: 1977.

URS Company. *Coal Train Assessment.* Summary Report URS No. 6108. Denver: 1976.

U.S. Coal Commission. *Report of the U.S. Coal Commission.* Washington, D.C.: Government Printing Office, 1925.

U.S. Commission on Civil Rights. *The Navajo Nation: An American Colony.* Washington, D.C.: 1975.

U.S. Congress. House. Committee on Education and Labor. *Black Lung Benefits Provisions of the Federal Coal Mine Health and Safety Act.* 95th Cong., 1st Sess., 14-17, 21 March 1977.

U.S. Congress. Office of Technology Assessment. *The Direct Use of Coal; Prospects and Problems of Production and Combustion.* Washington, D.C.: Government Printing Office, 1979.

U.S. Congress. Senate. Committee on Interstate Commerce. *Conditions in the Coalfields of Pennsylvania, West Virginia, and Ohio: Hearings on Senate Resolution 105.* 17th Cong., 1st Sess. 1928.

U.S. Congress. Senate. Permanent Subcommittee on Investigation of the Committee on Governmental Affairs. *The Coal Industry: Problems and Prospects; A Background Study.* Prepared by the Environment and Natural Resources Policy Division, Congressional Research Service, Library of Congress. 95th Congress, 2nd Sess. Washington, D.C.: Government Printing Office, 1978.

U.S. Department of Commerce. Bureau of the Census. *Housing Units Authorized by Building Permits and Public Contracts, May 1979.* Construction Reports C40-79-5. Washington, D.C.: Government Printing Office, 1979.

U.S. Department of Energy. *Energy Impact Assistance Study — Energy Project Data.* Washington, D.C.: October 1978.

U.S. Department of Energy. *Regional Profile: Energy Impacted Communities — Region VIII.* Oak Ridge, Tennessee: USDOE Technical Information Center, 1979.

U.S. Department of the Interior. *A Medical Survey of the Bituminous-Coal Industry.* Report of the Coal Mines Administration. Washington, D.C.: Government Printing Office, 1947.

U.S. Water Resources Council. *The Nation's Water Resources — 1975-2000,* Second National Water Assessment, Vol. 1. Washington, D.C.: Government Printing Office, 1978.

Watkins, Damon D. *Keeping the Home Fires Burning: A Book About the Coal Miner.* Ohio Company, 1937.

West Virginia Housing Development Fund. *Statewide Housing Needs Analysis for the State of West Virginia.*

Witt, Matt, with Earl Dotter (photographer). *In Our Blood; Four Coal Mining Families.* New Market, Tennessee: Highlander Researcher and Education Center, 1979.

Witt, Matt. "Medical Clinics: UMWA Families Organize to Breath New Life Into Coalfield Health Care." *United Mine Worker's Journal* No. 8 (April 16-30, 1975): 10-18.

Staff

Executive Director
Michael S. Koleda*

Executive Editor
L. Erick Kanter*

Editor
Joseph Foote

Writers
Lisa Berger
Claude E. Owre
Nada Samuels*

Advisers
James M. Childress*
William P. Hobgood
M. Blane Michael
Donald E. Pierce
Allan Sherman
D. Nathan Sumner
J. Allen Wampler*
Michael G. Willingham*
Bette P. Zabell*

Production Coordinator
Elizabeth S. Fitzsimmons*

Production Assistants
Louise M. Burris*
Marjorie M. Clayman
Venita Gibson*
Deborah S. Heath*
Carol A. Rogers
Cecelia E. Roots*
Rose Ann Tilton
Kathryn L. Wooten*

Field Survey Director
Douglas G. McConnell

Field Coordinator
Melissa A. Wolford*

Field Team Members
Catherine M. Allen
Jean E. Baldridge
Lisa Berger
R. Paul Brown
Jane Marie Drain
Nan E. Elliot
Mark B. Farber
James A. Little
Ronald McMahan
Eric Margolis
Mitchell Menzer*
Deborah G. Nordwall
Judy Rabinovitz
Kathryn Utrup
Michael Williams

Special Consultants
Barry O'Connell
William B. Walker
Ralph R. Widner
David C. Williams

Administration
Loraine C. Halsey*
Nancy L. Creason*
Christine M. Scoby*
Ellen P. Vancko*

Principal Photographers
Jack Corn
Kenneth Murray
Martha M. Simmons
Theodore M. Wathen

Special Advisers on Photography
Russell Lee
Robert Moeser

Design Liaison
Richard C. Reno
Kristina M. Jorgensen

Design and Production
William J. Kircher & Associates, Inc.
Heritage Arts, Inc.

Photography Assistant
Dina Rasor

Photographic Production
Manuel Gomez

Printing Liaison
William E. Kusterbeck
Ophelia H. Brown

*Members of Commission staff.

MARION

Scenes from the coalfields: Top left: A mobile home park near Gillette, Wyo. (Murray) Center far left: Removing overburden at a Consoldiation Coal Co. surface mine in Fulton County, Ill. (Wathen) Bottom left: Working underground at the Alabama By-Products Corp. Mary Lee #1 Mine near Jasper, Ala. (Murray) Center left: Gary, W. Va. (Corn) Top right: Colorado Westmoreland, Inc.'s Orchard Valley Mines outside Paonia, Colo. (Corn) Center right: Beth-Elkhorn Corp. Mine #131, Van, W. Va. (Corn) Bottom right: Loading coal in Harlan County, Ky. (Corn)

Scenes from the coalfields: Top left: Miners at Alabama By-Products Corp. Mary Lee #1 Mine near Jasper, Ala. (Murray) Bottom left: Cleaning plant, stock pile and railroad cars viewed from coal storage silo at Island Creek Coal Co., Inc., Providence #1 Mine in western Kentucky (Wathen). Center top: Jones & Laughlin Steel Corp. preparation plant, Vestaburg, Pa. (Wathen) Center middle: "Wee-Mah-Tuk" recreational area in Fulton County, Ill., built on a reclaimed surface mine. (Wathen). Center bottom: Houses and coal cars in Lyburn, W. Va. (Corn)
Right: Miners at Beth-Elkhorn Corp. Mine #131, Van, W. Va. (Corn).